D1175071

INORGANIC
SYNTHESES

Volume XIII

●●●

Editorial Board

FRED BASOLO   *Northwestern University*
HOWARD C. CLARK   *University of Western Ontario*
W. A. G. GRAHAM   *University of Alberta*
M. FREDERICK HAWTHORNE   *University of California (Riverside)*
RICHARD H. HOLM   *Massachusetts Institute of Technology*
ALAN G. MacDIARMID   *University of Pennsylvania*
GEORGE W. PARSHALL   *E. I. du Pont de Neumours and Company*
JOHN K. RUFF   *University of Georgia*
AARON WOLD   *Brown University*

Secretary, Editorial Board

STANLEY KIRSCHNER   *Wayne State University*

International Associates

E. O. FISCHER   *Technische Hochschule (Munich)*
JACK LEWIS   *University College (London)*
LAMBERTO MALATESTA   *University of Milan*
GEOFFREY WILKINSON   *Imperial College of Science
and Technology (London)*

Editor-in-Chief

F. A. COTTON
Professor, Department of Chemistry
Massachusetts Institute of Technology
Cambridge, Massachusetts

●●●●●●●●●●●●●●●●●●●●●●●●●●●●●●●●●●●●●●

INORGANIC
SYNTHESES

Volume XIII

McGRAW-HILL BOOK COMPANY

New York   St. Louis   San Francisco   Düsseldorf
Johannesburg   Kuala Lumpur   London   Mexico
Montreal   New Delhi   Panama   Rio de Janeiro
Singapore   Sydney   Toronto

CENTRAL METHODIST COLLEGE LIBRARY
FAYETTE, MISSOURI 65248

INORGANIC SYNTHESES, VOLUME XIII

Copyright © 1972 by McGraw-Hill, Inc. All Rights Reserved.
Printed in the United States of America. No part of this publication
may be reproduced, stored in a retrieval system, or transmitted,
in any form or by any means, electronic, mechanical, photocopying,
recording, or otherwise, without the prior written permission of
the publisher. *Library of Congress Catalog Card Number* 39-23015

07-013208-9

1234567890 MAMM 765432

To **JANET D. SCOTT**

*in appreciation of her many contributions
to* INORGANIC SYNTHESES *in the areas of
nomenclature and indexing.*

CONTENTS

x   *Contents*

PREFACE

In Volume XIII, detailed and checked synthetic procedures are presented for 120 compounds. These compounds are of three main classifications. First there are a number of compounds of the transition elements. Many of these are classical or Werner complexes and a few are organometallic compounds. There are also a few procedures dealing with the nonclassical compounds involving metal atom clusters or very strong multiple metal-metal bonds. A few more words about one of the latter will be found below. The second large group of compounds are those of nontransition elements. In the present volume this group is a rather diverse array, not strongly organized around any principal theme.

Although they are transition-metal compounds, I have chosen to place a group of complexes involving low-valent metals coordinated by various phosphorus ligands in a separate category, since they have an exceptional degree of both practical and fundamental interests. It is only relatively recently that the ability of many other ligands besides carbon monoxide, and most particularly various X_3P-type ligands, to stabilize metals in zero or other low formal oxidation states has been generally appreciated. It is now clear, however, that aryl phosphines and phosphites in general have this ability, and a number of compounds have been prepared. Many of these have shown promise as catalysts and are of interest, therefore, industrially as well as academically.

I should like to explain the background of the photograph on the dust jacket of this volume. It shows a postage stamp issued in 1969 by the Soviet Union. One of these stamps was presented to me in September 1969, at the Stony Brook meeting of the International Union of Crystallography by Professor N. A. Porai-Koshits. In the background is the N. S. Kurnikoff Institute where the chemistry of transition metals, particularly the heavier ones, is actively investigated. The legend at the upper right-hand corner says "Fiftieth Anniversary of the N. S. Kurnikoff Institute of the Union of Soviet Socialist Republic." On the left side of the stamp is a ball-and-stick model of an $Re_2X_8^{2-}$ (X = Cl, Br, etc.) ion. The chloro ion, $Re_2Cl_8^{2-}$, was first prepared in the N. S. Kurnikoff Institute and was reported by Russian workers in the early 1950s, albeit incorrectly characterized as a Re(II) compound, e.g., K_2ReCl_4. [See, for example, V. G. Kronev and S. M. Bondkin, *Khim. Redkikh. Elemetov, Akad. Nauk SSSR*, **I**, 40 (1954).] It was not until some 10 years later as part of the general renaissance of interest in rhenium chemistry that structural investigations were carried out on these compounds. In my own laboratory the same compounds were discovered independently and assigned the (correct) stoichiometry $Re_2Cl_8^{2-}$. Both in my own laboratory and at the N. S. Kurnikoff Institute x-ray investigations were undertaken which showed the structure to be that depicted on the stamp. A great deal of further chemical study of this and related species has been carried out, and it is now agreed that $Re_2X_8^{2-}$ is the proper stoichiometry and it has been shown that this species contains a quadruple bond.*

Inorganic Syntheses, Inc., is a nonprofit organization whose purpose is to help inorganic chemists with synthesis problems by providing detailed checked procedures for important compounds. It is hoped that the procedures which are published in these volumes will be sufficiently detailed and foolproof that even one new to the synthesis of compounds in a particular field

* For a recent review cf. F. A. Cotton, *Accts. Chem. Res.*, **2**, 240 (1969).

and unfamiliar with any special techniques or requirements in that field will be able to carry out the preparation successfully the first time. It is clear that an effort of this kind can be successful only as a result of the generous cooperation of many persons. It is important that people familiar with the synthetic procedures for important compounds in various fields take the initiative in preparing manuscripts suitable for INORGANIC SYNTHESES, or in seeing that others in the field do so. It is also essential that many people be willing to devote the time and attention necessary to check these procedures and work out in cooperation with the original authors any such modifications that appear necessary. I wish to acknowledge here my debt to many people who have assisted me in these ways in preparing Volume XIII. I would urge others in the inorganic field to keep INORGANIC SYNTHESES in mind and assist or stimulate the flow of suitable manuscripts. There follows a notice to contributors which describes the procedure for preparing and submitting such manuscripts.

Finally I should like to thank many members of Inorganic Syntheses, Inc., for their generous help and advice. I thank Dr. E. L. Muetterties and Professor G. Wilkinson for generous efforts in helping sort out some perplexing problems concerning conflicts, overlaps, and checking difficulties. I am especially grateful to Professors A. Wold, W. C. Fernelius, S. Kirschner, and S. Y. Tyree, who relieved me of the final burdens of preparing the manuscript for the printer when I became swamped by other commitments.

F. A. Cotton

NOTICE TO CONTRIBUTORS

The INORGANIC SYNTHESES series is published to provide all users of inorganic substances with detailed and foolproof procedures for the preparation of important and timely compounds. Thus the series is the concern of the entire scientific community. The Editorial Board hopes that all chemists will share in the responsibility of producing INORGANIC SYNTHESES by offering their advice and assistance both in the formulation and laboratory evaluation of outstanding syntheses. Help of this type will be invaluable in achieving excellence and pertinence to current scientific interests.

There is no rigid definition of what constitutes a suitable synthesis. The major criterion by which syntheses are judged is the potential value to the scientific community. An ideal synthesis is one which presents a new or revised experimental procedure applicable to a variety of related compounds, at least one of which is critically important in current research. However, syntheses of individual compounds that are of interest or importance are also acceptable.

The Editorial Board lists the following criteria of content for submitted manuscripts. Style should conform with that of previous volumes of INORGANIC SYNTHESES. The *Introduction* should include a concise and critical summary of the available procedures for synthesis of the product in question. It should also include an estimate of the time required for the synthesis, an indication of the importance and utility of the product, and an admonition if any potential hazards are associated with the procedure. The *Procedure* should present detailed and unambiguous laboratory directions and be written so that it anticipates possible mistakes and misunderstandings on the part of the person who attempts to duplicate the procedure. Any unusual equipment or procedure should be clearly described. Line drawings should be included when they can be helpful. All safety measures should be clearly stated. *Sources of unusual starting materials must be given,* and, if possible, minimal standards of purity of reagents and solvents should be stated. The scale should be reasonable for normal laboratory operation, and any problems involved in scaling the procedure either up or down should be discussed. The criteria for judging the purity of the final product should be clearly delineated. The section on *Properties* should list and discuss those physical and chemical characteristics that are relevant to judging the purity of the product and to permitting its handling and use in an intelligent manner. Under *References,* all pertinent literature citations should be listed in order.

The Editorial Board determines whether submitted syntheses meet the general specifications outlined above. Every synthesis must be satisfactorily reproduced in a different laboratory other than that from which it was submitted.

Each manuscript should be submitted in duplicate to the Secretary of the Editorial Board, Professor Stanley Kirschner, Department of Chemistry, Wayne State University, Detroit, Michigan 48202, U.S.A. The manuscript should be typewritten in English. Nomenclature should be consistent and should follow the recommendations presented in "The Definitive Rules for Nomenclature of Inorganic Chemistry," *J. Am. Chem. Soc.,* **82,** 5523 (1960). Abbreviations should conform to those used in publications of the American Chemical Society, particularly *Inorganic Chemistry.*

INORGANIC SYNTHESES

Volume XIII

Chapter One

COMPOUNDS OF THE NONTRANSITION ELEMENTS

1. PERBROMIC ACID AND POTASSIUM PERBROMATE

$$BrO_3^- + F_2 + 2OH^- \rightarrow BrO_4^- + 2F^- + H_2O$$

Submitted by EVAN H. APPELMAN*
Checked by JOHN R. BRAND†

Perbromates are obtained by oxidation of bromates in aqueous solution. Electrolytic oxidation, the best method of preparing perchlorates and periodates, is not very satisfactory in the case of perbromates. The most practical synthesis involves oxidation with elemental fluorine in sodium hydroxide solution. The procedure is somewhat hazardous, and the experimenter should familiarize himself with the precautions necessary for safely handling fluorine before undertaking the synthesis.[1]

After the oxidation, the sodium, fluoride, and excess bromate must be removed to obtain a pure product. The bulk of the bromate and fluoride are precipitated with barium, and the sodium is removed with cation exchange resin. After concen-

* Argonne National Laboratory, Argonne, Ill. 60439. Work performed under the auspices of the U.S. Atomic Energy Commission.
† Kansas State Teachers College, Emporia, Kans. 66801.

1

trating the solution, the remainder of the bromate is precipitated as silver bromate and the remaining fluoride as calcium fluoride. Ion exchange gives a solution of perbromic acid, which can be neutralized with the appropriate base to yield alkali or alkaline-earth salts. The perbromates of potassium, rubidium, and cesium have fairly low solubilities and can be readily prepared in a pure form.

To avoid pickup of silica, solutions containing appreciable amounts of fluoride and/or base should not be handled in glass. Polyethylene and polypropylene are satisfactory at room temperature, but heating must be done in Teflon vessels. Both TFE Teflon and the less expensive FEP (fluorinated ethylene propylene) may be utilized, but the latter must be used with greater care to prevent overheating. Magnetic stirring bars should be Teflon coated. Coarse porous polyethylene filters are available (Porex Materials Corp.), but for filtrations requiring fine porosity, Teflon filter disks (Chemplast Inc.) should be used on polypropylene or polyethylene Büchner funnels.

Assays of bromate and perbromate concentrations are required during the procedure. Bromate concentrations that are at least comparable to the perbromate may be determined iodometrically by reaction with sodium iodide in acid solution containing molybdate, followed by titration with standardized thiosulfate. After reduction of the bromate the solution should be *ca.* 0.1 M each in H^+ and in free iodide ion. Perchloric, hydrochloric, or sulfuric acids may be used. The molybdenum(VI) concentration should be *ca.* 10^{-3} M.

To determine the perbromate concentration of a solution containing both bromate and perbromate, enough 48% hydrogen bromide is added to provide an excess of 0.5–1.5 M hydrogen bromide after reduction of the bromate to tribromide ion, Br_3^-, and neutralization of any base present. (For this purpose sodium fluoride constitutes a base.) Pure nitrogen or argon is bubbled through the solution until all color is gone. The solution is then diluted with at least five times its volume of saturated hydrogen bromide. (The hydrogen bromide concentration after reaction must be at least 11.5 M.) The vessel containing the solution is stoppered with a glass or Teflon stopper and is allowed to stand for 10 minutes. Its contents are then transferred rapidly and quantitatively to about 20 times their volume of water containing slightly more than enough sodium dihydrogen phosphate to neutralize the hydrogen bromide, and containing enough sodium iodide to provide a 0.1 M excess. The triiodide formed is then titrated with standardized thiosulfate. If the bromate concentration of the original solution is negligible, the addition of 48% hydrogen bromide and subsequent gas flushing may be omitted, and saturated hydrogen bromide may be added directly.

Inasmuch as the labor involved is nearly independent of the size of the preparation, the following procedure is designed on a fairly large scale. It may, of course, be scaled down to suit the needs and available equipment of the laboratory using it. If the procedure is followed as written, a centrifuge capable of holding 1-l. polypropylene bottles will be very useful. The use of a smaller centrifuge, of course, will merely make the operations more time-consuming. Use of a rotary vacuum evaporator will expedite the concentration steps in the procedure. Either a batch- or continuous-feed unit may be employed, but it should be of one of the types that cannot contaminate the concentrate with grease or metal. Heat may be applied to the evaporator as fast as it can be absorbed without causing the solution to boil.

A heavy-duty magnetic stirrer, such as Cole-Parmer model 4817, is needed during the fluorination.

Procedure

Equip a $\frac{1}{2}$-lb. cylinder of fluorine with a pressure-reducing regulator especially designed for fluorine service. Mount it securely in a well-ventilated fume hood. If the hood uses fiberglass exhaust filters, they should be of the "perchloric acid type," which contains no organic binders.

Attach a Monel or brass needle valve to the low-pressure side of the regulator. It should either have Teflon packing or be of the packless type with soldered-on bellows. Attach a 2-ft. length of $\frac{1}{4}$-in.-o.d. copper tubing to the valve through either a flare fitting or a swage-type tube fitting. Intermediate fittings on the low-pressure side of the regulator may be made of Monel, brass, or aluminum. Pipe fittings should be sealed only with Teflon thread tape. Permanent connections may be made with silver solder.

The copper tube should be bent so that it extends to a point about 1 ft. away from the cylinder and at least 2 ft. above the floor of the fume hood. The last inch of its length should be bent to point straight down. A 1-ft. length of TFE Teflon tubing, $\frac{3}{8}$ in. o.d. $\times$ $\frac{1}{4}$ in. i.d., is then forced over the end of the copper tube.

Make up 900 ml. of 5 M sodium hydroxide in a 2-l. Teflon FEP narrow-mouthed bottle (Nalge Co.). Add 200 g. sodium bromate and introduce a large Teflon-clad magnetic stirring bar. Stir the mixture for 20 minutes with a heavy-duty magnetic stirrer. Only a portion of the bromate will dissolve. Surround the bottle with water and crushed ice in a 4-l. beaker.

Start a gentle flow of fluorine from the cylinder and raise the beaker and bottle to immerse the Teflon tubing in the alkaline bromate solution. The end of the Teflon tube should be just above the stirring bar. Position the heavy-duty stirrer beneath the beaker and stir vigorously enough to keep the solid sodium bromate distributed throughout the solution. The fluorine flow may now be made very rapid, provided undue spattering does not result. However, care must be taken to maintain the ice bath around the Teflon bottle. Even with this cooling, the temperature of the solution may approach its boiling point, and occasional small, but noisy, detonations may occur in the vapor above the solution. *The reaction must never be left unattended!*

If a flame appears in the bottle, the needle valve should be shut momentarily to quench it. If it is necessary to terminate the fluorine flow for more than a moment, the tube should be withdrawn from the bottle and cleared of any liquid with a burst of fluorine.*

The absorption of fluorine may be monitored by the decrease in the pressure in the cylinder. When most of the alkali has been consumed, the flow rate should be reduced. The approach to neutrality will be indicated by the appearance of fumes at the mouth of the bottle. If fluorine is added substantially beyond the neutral point, the solution will turn yellow. However, perbromate is formed only in alkaline medium.

When the solution is nearly neutral, add 300 ml. 50% sodium hydroxide and 80 g. sodium bromate. Again introduce fluorine until the solution approaches neutrality. Then add 400 ml. 50% sodium hydroxide and 100 g. sodium bromate. Once more pass in fluorine until the solution is almost neutral. This completes the fluorination step.

In the following purification procedure, glass vessels must be avoided except in rotary evaporators or where especially designated. "Washing" of a precipitate or resin will imply washing until negligible additional amounts of perbromate are removed. Washes are always combined with the original filtrate, eluate, or supernatant solution for the next step.

Using a Teflon delivery tube, pass a vigorous stream of pure nitrogen or argon through the reaction mixture for 5 minutes to expel oxygen fluoride and remove unreacted fluorine from the space above the mixture. Cool the mixture to room temperature or below and stir for 20 minutes. Transfer to a polypropylene centrifuge bottle. Centrifuge, wash, and discard the precipitated sodium fluoride.

Add slowly, with stirring, 300 g. of *anhydrous* barium hydroxide for each liter of solution. Continue stirring until the mixture

---

* The checker prefers the use of the nitrogen purge assembly described in reference 1 for flushing out the line after use.

cools back to room temperature, but for not less than 1 hour. Centrifuge, wash, and discard precipitate.

Add enough analytical grade Dowex 50X8 cation exchange resin, 20–50 mesh, in the hydrogen form, to the solution to make it at least 0.05 M in acid. Filter with suction through a coarse polyethylene or Teflon filter and wash the resin. Neutralize the filtrate with calcium carbonate and concentrate to *ca.* 400 ml., either in a rotary evaporator or in a Teflon beaker under a heat lamp.

If an evaporator was used, transfer the concentrate to a Teflon beaker. Assay the bromate concentration and add with stirring enough saturated silver fluoride to provide a 0.1–0.2 M excess after precipitation of silver bromate. Centrifuge. Wash the precipitate with 0.1 M silver fluoride and discard it. Pass the supernatant solution and washings under suction through an extra-fine Teflon filter disk on a polyethylene Büchner funnel.

To the filtrate add gradually with stirring a 10–20% excess of calcium hydroxide over the amount needed to precipitate all the added fluoride as calcium fluoride. Continue to stir for at least one hour. Centrifuge, wash, and discard precipitate.

Again acidify the solution with Dowex 50 and filter as before. Neutralize the filtrate with calcium hydroxide and add enough excess to saturate the solution. Add 20 mg. of diatomaceous earth filter aid per liter of solution. In a fine sintered-glass filtering funnel, slurry 100 mg. of the filter aid per square centimeter of filter surface. Suck most of the water from the slurry through the funnel, but before the slurry is quite dry, filter the perbromate solution through it. Wash with saturated calcium hydroxide.

Pack an ion exchange column with analytical grade Dowex 50X8, 50–100 mesh, in the hydrogen form. The column should be 6–8 cm. i.d. and high enough to provide four equivalents of exchange capacity per mole of perbromate in the solution. Pass the perbromate solution through the column and wash through with water. Using a heat lamp or rotary evaporator, concentrate the eluate to obtain about 250 ml. of 4 M perbromic acid, or a *ca.* 10% yield on the basis of the fluorine consumed.

If potassium perbromate is the desired product, neutralize the solution potentiometrically with 4 M potassium hydroxide, using 0.1 M potassium hydroxide to reach the precise end point. It is advisable to keep a little of the perbromic acid solution in reserve in case the end point is overshot.   Glassware may be used for this and subsequent steps.

Heat the potassium perbromate slurry to 100° and add enough water to bring all the solid into solution at this temperature. Allow to cool gradually to room temperature.   Then chill in an ice bath for an hour and decant the supernatant solution. Redissolve the solid in a minimum of water at 100° and again cool, chill, and decant.   Dry the solid at 100°.   Transfer to an agate mortar, crush, and dry to constant weight at 110° in vacuum.   The yield of potassium perbromate is 80% of the perbromic acid taken.   If the recrystallization step is omitted, the yield is 90%.

Properties[2]

Perbromic acid is a strong monobasic acid.   Its aqueous solutions are stable up to about 6 M (55% $HBrO_4$), even at 100°. Fairly concentrated solutions may develop a yellow bromine color from the decomposition of traces of bromate ion and hypobromous acid.   If a 6 M perbromic acid solution is allowed to stand for several months, the bromate and hypobromite will have all decomposed, and the resulting bromine can be flushed out with pure nitrogen, leaving a colorless solution.

Above 6 M, perbromic acid solutions tend to be erratically unstable, although the decomposition is not explosive.   Concentration in vacuum at room temperature produces an azeotrope consisting of about 80% perbromic acid (*ca.* 12 M), which usually decomposes during or shortly after preparation.   Molecular distillation of this azeotrope is possible if heat is applied rapidly in high vacuum.

The bromate–perbromate electrode potential is about 1.76 volts in acid solution,[3] making perbromic acid a potent oxidant. However, dilute solutions react sluggishly at room temperature. Bromide and iodide are oxidized slowly and chloride not at all. The 6 M acid attacks stainless steel at room temperature, and at 100° it oxidizes chloride ion to chlorine, Cr(III) to Cr(VI), Mn(II) to MnO_2, and Ce(III) to Ce(IV) in nitrate solution. The 12-M acid is a vigorous oxidizing agent even at room temperature.

Pure potassium perbromate is stable up to 275°, at which temperature it decomposes to potassium bromate. The impure product may undergo partial decomposition at lower temperatures.

Analytical[2]

Perbromic acid and perbromates are most readily assayed by determination of their oxidizing power after reduction with hydrogen bromide, as described earlier in this article. Traces of fluoride in the acid or salts may be determined potentiometrically, using a fluoride-sensitive electrode (Orion Research, Inc.) and an expanded-scale pH meter. Acid or alkaline solutions should be neutralized or buffered with acetic acid and sodium acetate before the determination. The electrode response should be calibrated against similar solutions of known fluoride content.

Such lower bromine oxidation states as Br_2, HOBr, $HBrO_2$, and BrO_3^- can be estimated by conversion to tribromide in 0.5 M hydrogen bromide. The tribromide can be determined spectrophotometrically at 275 nm. At this wavelength, perbromic acid has an extinction coefficient of only about 5.8 l./mole-cm., whereas tribromide has an extinction coefficient of about 3.5×10^4. The latter is reduced to an apparent value of about 3.1×10^4 by the incomplete formation of tribromide. Inasmuch as large amounts of perbromate will slowly oxidize $0.5\ M$ hydrogen bromide, the absorption should be followed as a function of time and extrapolated back to the time of mixing.

References

1. "Matheson Gas Data Book," Matheson Co., Inc., East Rutherford, N.J., 1961.
2. E. H. Appelman, *Inorg. Chem.*, **8**, 223 (1969).
3. G. K. Johnson, P. N. Smith, E. H. Appelman, and W. N. Hubbard, *Inorg. Chem.*, **9**, 119 (1970).

2. α-SULFANURIC CHLORIDE—CYCLIC TRIMER
(*1,3,5-Trichloro-1H,3H,5H-1,3,5,2,4,6-trithiatriazine 1,3,5-trioxide*)

$$H_2NSO_3H + 2PCl_5 \xrightarrow{CCl_4} Cl_3P{=}NSO_2Cl + POCl_3 + 3HCl$$
$$3Cl_3P{=}NSO_2Cl \xrightarrow{127–137°} (NSOCl)_3 + 3POCl_3$$

Submitted by THERALD MOELLER,* TIAO-HSU CHANG,† AKIRA OUCHI,‡
ANTONIO VANDI,§ and AMEDEO FAILLI¶
Checked by W. E. HILL‖

The cyclic sulfanuric chlorides, $(NSOCl)_n$, are of interest because they are isoelectronic with the cyclic phosphonitrilic chlorides, $(NPCl_2)_n$. Although the formation of a variety of substances with $n = 3$ or more appears reasonable, only the cyclic trimers have been isolated.[1] At least three, apparently conformational, isomers of composition $(NSOCl)_3$ have been

* Arizona State University, Tempe. Ariz. 85281.
† National Taiwan University, Taipei, Taiwan.
‡ College of General Education, University of Tokyo, Komaba, Meguro-ku, Tokyo, Japan.
§ Naval Ordnance Station, Indian Head, Md. 20640.
¶ University of British Columbia, Vancouver, British Columbia, Canada.
‖ Rohm and Haas Company, Redstone Research Laboratories, Huntsville, Ala. 35807.

reported,[1,2] but of these only the α isomer is thermodynamically stable at room temperature. The trimeric sulfanuric chlorides have been prepared by the thermal decomposition of (trichlorophosphoranylidene)sulfamoyl chloride,[1,2] by the ammonolysis of sulfuryl chloride in admixture with thionyl chloride,[3] and by the oxidation of trithiazyl chloride, $(NSCl)_3$, with sulfur(VI) oxide.[4]

The first of these procedures is the most convenient and gives the highest yields. (Trichlorophosphoranylidene)sulfamoyl chloride is readily obtained in excellent yield by the reaction of sulfamic acid with phosphorus(V) chloride,[1,2,5] and its pyrolytic decomposition is best effected at 127–137° under a pressure of 8–9 mm. Hg and with a slow purge of dry air or nitrogen. Both α- and β-sulfanuric chloride result, but conversion of the latter to the former is essentially complete in the procedure as given below. The β isomer can be obtained by immediately extracting the product of the thermal decomposition reaction with *n*-heptane, removing the α-isomer by crystallization at 0°, evaporating the filtrate *in vacuo*, and subliming *in vacuo* at room temperature.[2]

Inasmuch as phosphorus(V) chloride, phosphorus(V) oxytrichloride, and (trichlorophosphoranylidene)sulfamoyl chloride are all highly sensitive to moisture, operations that cannot be carried out in closed systems must be done in a gloved dry-box.

■ *Caution. Contact between any of these substances and water in a closed system can lead to a dangerous explosion!*

Procedure

A. (TRICHLOROPHOSPHORANYLIDENE)SULFAMOYL CHLORIDE

To a dry, 1000-ml., round-bottomed flask with a single standard-taper opening are added 110 g. (1.1 moles) of finely pulverized sulfamic acid [dried at 100° and cooled *in vacuo* over phosphorus(V) oxide], 450 g. (2.2 moles) of phosphorus(V) chloride, and 60 ml. of freshly distilled carbon tetrachloride. A reflux condenser, to which a drying tube containing calcium chloride is attached, is immediately fitted to the flask. The flask and its contents are then heated in an oil bath at 80–85° for 15 hours. The reaction mixture is then cooled to 0° and filtered at that temperature in the gloved dry-box, using a medium-frit, sintered-glass vacuum filter.*

The filtrate is evaporated in the hood at 85° and 1 mm. Hg to remove both carbon tetrachloride and phosphorus(V) oxytrichloride. The transparent, nearly colorless, oily residue is stored overnight in the refrigerator (at 0–5°) to crystallize. The solid product is separated from oily contaminants by pouring through a coarse-frit, sintered-glass crucible as rapidly as possible after transfer of the container from the refrigerator to the gloved dry-box.* Dry nitrogen is pulled through the crystals on the filter until the solid appears dry. The solid is transferred from the filter to a dry vial and stored under dry nitrogen. The filtrate is returned to the refrigerator and the entire operation repeated until no additional crystalline product can be recovered.† M.P., 33°. Yield, based upon phosphorus(V) chloride, is 180 g. (67%).

B. α-SULFANURIC CHLORIDE—CYCLIC TRIMER

One hundred twenty-five grams (0.5 mole) of (trichlorophosphoranylidene)sulfamoyl chloride is placed in a 300-ml., three-necked, round-bottomed flask, to which a gas inlet tube, a thermometer, and a Vigreux condenser are attached. To the top of the Vigreux column, a Liebig condenser is attached. This condenser is, in turn, attached to a receiver which is immersed in a Dry Ice–acetone mixture.‡ The outlet from the

---

* Ideally, all operations should be carried out in a cold room maintained below 5°.
† ■ *Caution.* *The malodorous final filtrate should be disposed of out-of-doors. It may react explosively with water.*
‡ ■ *Note.* *This distillate should be disposed of in the same fashion as the filtrate from Sec. A.*

receiver leads through a similarly cooled trap to a vacuum pump. While a current of *dry* air or nitrogen is passed through the reaction vessel, the latter is heated at 127° for one hour and then at 137° for an additional hour. The internal pressure of the system is maintained at 8–9 mm. Hg.

The mixture remaining in the three-necked flask is stored overnight at 0°. The solid product is then freed of oily contaminants by spreading on clay plates, which are stored in a desiccator. Finally, the solid is washed with 100 ml. of ice water and then dried in a desiccator over phosphorus(V) oxide. The crude product so obtained is recrystallized twice from 10–15 times its weight of n-heptane which has been previously dried over sodium wire. M.P., 144–145°. Yield, based upon (trichlorophosphoranylidene)sulfamoyl chloride, is *ca.* 12 g. (17%). *Anal.* Calcd. for $N_3S_3O_3Cl_3$: N, 14.36; S, 32.87; Cl, 36.36; mol. wt., 292.57. Found: N, 14.51; S, 32.44; Cl, 36.84; mol. wt., 300.

Properties

α-Sulfanuric chloride—cyclic trimer is obtained as transparent, colorless, prismatic crystals, melting at 144–145°. It is insoluble in and only slowly hydrolyzed by cold water. Its solubility at 25° in grams per 100 g. of solvent has been reported[2] as: C_6H_6, 22.50; CH_3CN, 13.15; CS_2, 4.10; CCl_4, 2.95; petroleum ether (90–110°), 2.32; C_6H_{12}, 1.63; n-C_7H_{16}, 1.56. Petroleum ether (90–110°), cyclohexane, and n-heptane are suitable solvents for recrystallization. The β isomer is much more soluble in each of these solvents.

The infrared spectrum shows strong sulfur-oxygen stretching modes at 1344 and 1110 cm.$^{-1}$. Strong bands at 700 and 665 cm.$^{-1}$ are associated with sulfur-nitrogen stretching modes. The spectrum of the β isomer is nearly identical. α-Sulfanuric chloride—cyclic trimer crystallizes in the orthorhombic system ($a = 7.552$, $b = 11.540$, $c = 10.078$ A.), space group *Pnma* (D_{2h}^{16}), with four molecules to the unit cell.[7]   Mean bond lengths in the molecule are: S—N, 1.571 ± 0.004; S—Cl, 2.003 ± 0.003; and S—O, 1.407 ± 0.007 A. Mean bond angles are: NSN, 112.8 ± 0.4; SNS, 122.0 ± 0.4; NSO, 111.9 ± 0.35; NSCl, 106.3 ± 0.3; OSCl, 107.9 ± 0.35°. The nitrogen and sulfur atoms alternate in a ring that has a slight chair-type conformation. The arrangement of bonds around each sulfur atom is roughly tetrahedral. The chlorine atoms are all axial, the oxygen atoms equatorial. The short, equal S—N bond distances suggest p_π-d_π electron-density delocalization within the ring. The dipole moments of the α and β isomers are, respectively, 3.88 and 1.91 debyes.[2] The β isomer may differ in the arrangements of some of the chlorine and oxygen atoms but probably not in the ring conformation.[2]

The α-sulfanuric chloride—cyclic trimer molecule undergoes ring cleavage when treated with strongly basic amines but gives two isomeric trisubstituted derivatives (m.p., 171–172 and 196–197°) when treated with the more weakly basic morpholine.[8] Ring cleavage in its reaction with benzene in the presence of aluminum chloride and triethylamine gives diphenyl sulfoxide. Reaction with potassium fluoride in acetonitrile, with a catalytic quantity of water present, yields a mixture of cis and trans sulfanuric fluoride-cyclic trimers, $N_3S_3O_3F_3$.[9,10] α-Sulfanuric chloride—cyclic trimer decomposes explosively at 285°.[11]

References

1. A. V. Kirsanov, *J. Gen. Chem. U.S.S.R.*, **22**, 93 (1952).
2. A. Vandi, T. Moeller, and T. L. Brown, *Inorg. Chem.*, **2**, 899 (1963).
3. M. Goehring, J. Heinke, H. Malz, and G. Roos, *Z. Anorg. Allgem. Chem.*, **273**, 200 (1953).
4. M. Goehring and H. Malz, *Z. Naturforsch.*, **9b**, 567 (1954).
5. M. Goehring, H. Hohenschutz, and R. Appel, *Z. Naturforsch.*, **9b**, 678 (1954).
6. M. Becke-Goehring and E. Fluck, *Inorganic Syntheses*, **8**, 105 (1966).
7. A. C. Hazell, G. A. Wiegers, and A. Vos, *Acta Cryst.*, **20**, 186 (1966).
8. A. Failli, M. A. Kresge, C. W. Allen, and T. Moeller, *Inorg. Nucl. Chem. Letters*, **2**, 165 (1966).
9. F. Seel and G. Simon, *Z. Naturforsch.*, **19b**, 354 (1964).
10. T. Moeller and A. Ouchi, *J. Inorg. Nucl. Chem.*, **28**, 2147 (1966).
11. R. L. McKenney, Jr., and N. R. Fetter, *J. Inorg. Nucl. Chem.*, **30**, 2927 (1958).

3. ARSINE AND ARSINE-d_3

$$Na_3As + 3H_2O(D_2O) \rightarrow AsH_3(AsD_3) + 3NaOH(NaOD)$$

Submitted by JOHN E. DRAKE* and CHRIS RIDDLE*
Checked by JOHN R. WEBSTER† and WILLIAM L. JOLLY†

Arsine, AsH_3, may be prepared conveniently by the potassium tetrahydroborate reduction of arsenic(III) oxide,[1] but this method is not readily adaptable for the formation of arsine-d_3, AsD_3. The procedure described here allows the ready preparation of arsine and arsine-d_3 from inexpensive materials with excellent yields. Only the simplest, readily available, vacuum-line apparatus is required.‡ The sodium arsenide alloy Na_3As is formed *in situ* from its elements, and addition of water[2] or heavy water then gives the arsines. The products are of high purity and need only drying before use. The deuterium content of the arsine-d_3 is of the same order as that of the deuterium oxide used. The yield (based on Na_3As) is essentially quantitative, since without using any elaborate trapping techniques 85–90% of the arsenic is recovered as arsine.

■ *Caution.* *Arsine is a colorless, poisonous gas that requires careful handling. A suitable gas mask must be at hand in case an on-line fracture occurs during the preparation. The pump exhaust should lead to the open air or a fume hood.*

Procedure

The apparatus must be clean, dry, and sound. Sodium (7.5 g.) is cut into small pieces under oil, dried, and placed in a

---

* Department of Chemistry, University of Windsor, Windsor, Ontario, Canada.

† Inorganic Materials Research Division of the Lawrence Radiation Laboratory, Berkeley, Calif. 94720.

‡ The apparatus lends itself to several preparations, e.g., H_2S or D_2S from $Al_2S_3 + H_2O(D_2O)$ and H_2Se or D_2Se from Al_2Se_3[3] in which the alloys are formed in a fume hood and then transferred to the apparatus.

nickel crucible.   An excess of finely powdered arsenic (*ca.* 15 g.) is then placed around the sodium.   Since the alloy formation is exothermic, an insulating layer of glass wool is placed at the bottom of the round-bottomed, cylindrical reaction vessel. The crucible, with reactants, is placed on the glass wool to avoid direct contact with the vessel.

The assembly is completed as in Fig. 1, and the reaction vessel, U traps, and collection vessel are evacuated.   The system is flushed thoroughly with dry nitrogen or argon.   After a final evacuation, the taps to the pump are closed and nitrogen or argon added to a pressure of 50–100 mm.

The bottom of the reaction vessel is heated *very gently* with a semiluminous Bunsen flame which is removed at the *first signs* of sodium melting.   A rapid reaction follows immediately, and if the heating has been excessive the mixture glows red-hot. It is quite normal for a black cloud to rise up the reaction

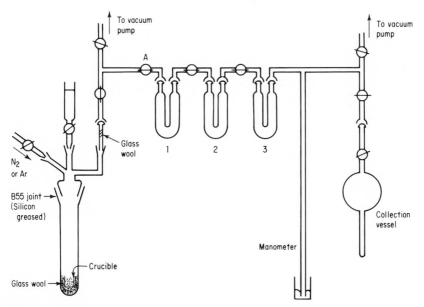

Fig. 1.   Apparatus for the preparation of arsine.

vessel. To ensure that no solid material enters the vacuum line, a loose-fitting plug of glass wool may be placed in the tube connecting the vacuum line to the reaction vessel whose dimensions (30 cm. long; 6.5 cm. o.d.) serve to minimize such a risk.

When the crucible has cooled down, the system is thoroughly evacuated, and water, or heavy water (*ca.* 10 ml.), allowed to drip slowly onto the alloy. The progress of the reaction can be followed by observing the manometer. When no more gas is formed, the products are condensed successively in the traps 1, 2, and 3 and in the collection vessel. The tap *A* is closed when the pressure drops below *ca.* 10 mm. to keep to a minimum the amount of water that distills over. The reaction vessel is evacuated, filled with N_2 or Ar, and removed to a fume hood.

Any noncondensible gas in the product is pumped off through the three U traps, and the crude product distilled into trap 1. Trap 2 is surrounded by a −111° slush bath (1-bromobutane or carbon disulfide) and the arsine distilled into the collection vessel. Typically, 2000 cc. (S.T.P.) or 0.095 moles of purified arsine is obtained (88% based on Na_3As).

Properties

The arsine may be stored in a glass vessel fitted with a greased tap. It has a vapor pressure of 35 mm. at −111.6°,[4] a melting point of −116.93°, and a boiling point of −62.48°.[5] Because of the 100% natural abundance of [75]As, the isotopic purity of arsine-d_3 can be readily established by mass spectroscopy. The infrared spectra (typically recorded at 50 mm. pressure in a 5-cm.-path-length cell) have strong absorptions at 2122, 1005, and 906 cm.$^{-1}$ for AsH_3 and at 1534, 714, and 660 cm.$^{-1}$ for AsD_3.[6]

References

1. J. E. Drake and W. L. Jolly, *Inorganic Syntheses*, **7**, 41 (1963).
2. See also R. E. Stroup, R. A. Oetjen, and E. E. Bell, *J. Opt. Soc. Am.*, **43**, 1096 (1953).

3. G. R. Waitkins and R. Shutt, *Inorganic Syntheses*, **2**, 183 (1946).
4. W. C. Johnson and A. Peckukas, *J. Am. Chem. Soc.*, **59**, 2065 (1937).
5. R. H. Sherman and W. F. Giauque, *ibid.*, **77**, 2154 (1955).
6. E. Lee and C. K. Wu, *Trans. Faraday Soc.*, **35**, 1366 (1939).

4. AMMONIUM CYANATE

$$(HNCO)_3 \rightarrow 3HNCO$$
$$HNCO + NH_3 \rightarrow NH_4NCO$$

Submitted by RODGER B. BAIRD,* and ROBERT P. PINNELL*
Checked by A. L. ALLRED,† D. V. STYNES,† and D. L. DIEMANTE†

Ammonium cyanate has the historical distinction of being the "inorganic" intermediate in Wohler's classical synthesis of "organic" urea.[1] Although based upon obvious metathetical reactions between ammonium chloride and silver cyanate or aqueous ammonia and lead cyanate, actual isolation of crystalline ammonium cyanate was not accomplished. This type of reaction has, however, found use in the preparation of urea-^{14}C-^{15}N via the intermediate ^{15}NH$_4$N^{14}CO.[2] Solid ammonium cyanate has been prepared commercially by flash evaporation of urea at 300–350°,[3] whereas aqueous solutions have been prepared by passage of natural gas and ammonia over alumina at 1000° followed by absorption of the effluent gas in aqueous ammonia.[4] Extension of the method used in preparing sodium and potassium cyanate, i.e., fusion of the appropriate alkali-metal carbonate with urea,[5] is obviously not feasible for the ammonium salt. The following procedure is based upon that of Waddington[6] and has been found to give a pure product in reasonably high yields. Preparation of the crude salt requires approximately 1–2 hours;

* Joint Science Department, Scripps, Pitzer, and Claremont Men's Colleges, Claremont, Calif. 91711. Financial support of the Research Corporation is gratefully acknowledged.

† Northwestern University, Evanston, Ill. 60201.

purification to analytical purity necessitates extraction for an additional 8–12 hours.

Procedure

The apparatus consists of a Pyrex tube, 20 mm. in diam., 400 mm. in length, equipped with a gas inlet from a source of dry nitrogen and an outlet of approximately 10 mm. in diam. The outlet leads directly into a 200-ml., three-necked flask equipped with a mechanical stirrer. A side neck of the flask holds a second gas inlet connected to a source of anhydrous ammonia and an outlet leading to a fume vent. Dry nitrogen is used both as a carrier for the isocyanic acid and as a readily available inert atmosphere. ■ *Caution. The preparation should be carried out in a fume hood because of the highly toxic character of isocyanic acid and its derivatives.*

At the upper end of the tube is placed 6.46 g. of recrystallized cyanuric acid (0.05 mole of trimer).* A volume of 100 ml. of anhydrous diethyl ether is poured into the flask and cooled to 0° in an ice bath. The gas inlet tubes should dip just below the surface of the ether while it is being stirred.

The cyanuric acid is heated slowly but uniformly, using either a gentle flame† or a tube furnace at a temperature of 380–400°. The acid sublimes easily and may condense in the cooler portions of the tube if heating is not uniform. Depolymerization proceeds readily, but extensive overheating of the material should be avoided because of decomposition and subsequent contamination of the product with ammonium cyanide.[7] The isocyanic acid formed in the depolymerization is conducted slowly into the stirred diethyl ether.‡

* Commercial material recrystallized from aqueous solution.

† The flame source of heat is more useful in controlling the rate of decomposition versus sublimation but, because of the presence of diethyl ether, requires increased emphasis upon carrying out the reaction in a well-ventilated area.

‡ Rapid cooling of the vapor should be avoided. At lower temperatures, appreciable amounts of cyanic acid, HOCN, are formed. This acid is unstable, polymerizing rapidly to mixtures of cyanuric acid and cyamelide. At 0°, cyanic acid exists as a liquid and the rate of polymerization may be explosive.[8]

After all of the cyanuric acid has been depolymerized, anhydrous ammonia is passed directly into the ethereal solution of isocyanic acid. Precipitation is immediate, and efficient stirring is necessary to prevent clogging of the inlet tube. Because of the voluminous nature of the precipitate, completion of the neutralization is difficult to ascertain and an obvious excess of ammonia should be passed into the mixture.

The white precipitate is filtered on a fritted-disk extraction thimble (EC porosity). Ammonium cyanide and cyanuric acid are removed from the product by overnight extraction with approximately 100 ml. of diethyl ether using a Soxhlet extractor. The precipitate in the thimble should be stirred several times during the extraction to effect maximum contact with the solvent. Yield is 4.70 g. of white powder (52%, based upon cyanuric acid). *Anal.* Calcd. for NH_4NCO: C, 20.00; H, 6.71; N, 46.65. Found: C, 19.93; H, 6.79; N, 46.59.

Properties

Ammonium cyanate is a white solid which crystallizes in the tetragonal system.[6] The solid rearranges primarily to urea upon heating or prolonged storage; the salt should be freshly prepared if used as an intermediate where the presence of urea is objectionable. It is extremely soluble in water, partially soluble in ethanol and chloroform, and insoluble in diethyl ether and benzene. Heating aqueous solutions of the salt causes extensive rearrangement.[9] Unfortunately, elemental analysis will not yield any information as to the extent of contamination by urea. Its presence can be determined qualitatively via the infrared spectrum of the solid mulled in mineral oil; urea exhibits strong absorptions at 1683 and 3456 cm.$^{-1}$, whereas ammonium cyanate absorbs radiation at 3160(s), 2190(s), 1334(m), and 640(s) cm.$^{-1}$.

References

1. F. Wohler, *Poggendorf's Annalen der Physik und Chemie,* **12,** 253 (1828). Reprinted in H. M. Leicester and H. S. Klickstein, "A Sourcebook in Chemistry," p. 309, McGraw-Hill Book Company, New York, 1952.

2. D. L. Williams and H. R. Ronzio, *J. Am. Chem. Soc.*, **74**, 2407 (1952).
3. L. G. Boatright (American Cyanamid Co.) U.S. Patent 2,712,491 (July 5, 1955); *C. A.*, **49**, 15189f (1955).
4. J. Kato, R. Iwanaga, and I. Hayashi (Ajinomoto Co., Inc.), Japanese Patent 6978 (June 13, 1960); *C. A.*, **55**, 18035e (1961).
5. A. Scattergood, *Inorganic Syntheses*, **2**, 87, 88 (1946).
6. T. C. Waddington, *J. Chem. Soc.*, **1959**, 2499.
7. G. Brauer, ed., "Handbook of Preparative Inorganic Chemistry," 2d ed., Vol. 1, p. 668, transl. by Scripta Technica, Inc., Academic Press Inc., New York, 1963.
8. E. M. Smolin and L. Rapoport, "*s*-Triazines and Derivatives," "The Chemistry of Heterocyclic Compounds," Vol. 13, p. 25, Interscience Publishers, Inc., New York, 1959.
9. J. Walker and F. J. Hambly, *J. Chem. Soc.*, **67**, 746 (1895).

5. PHOSPHORUS(III) ISOCYANATE
(*Phosphorus Triisocyanate*)

$$PCl_3 + 3AgNCO \rightarrow P(NCO)_3 + 3AgCl$$

Submitted by ROBERT E. ZOBEL* and ROBERT P. PINNELL*
Checked by MARY F. SWINIARSKI† and ROBERT R. HOLMES†

Phosphorus(III) isocyanate was first prepared by the reaction of phosphorus(III) chloride with silver isocyanate in warm benzene.[1] A later modification utilized phosphorus(III) iodide as a starting material with nitromethane as solvent.[2]

Less-expensive preparations, in terms of materials, have involved reactions between phosphorus(III) chloride and lithium cyanate, in benzene,[3] or sodium and potassium cyanates in polar solvents such as nitriles, ketones, nitroparaffins, and esters.[4] Liquid sulfur dioxide has also been found useful as a solvent for reaction between phosphorus(III) chloride and sodium cyanate.[5]

The synthesis from silver isocyanate and phosphorus(III) chloride has proved the most dependable and gives the highest

* Joint Science Department, Scripps, Pitzer, and Claremont Men's Colleges, Claremont, Calif. 91711. Financial support of the Research Corporation is gratefully acknowledged.

† Department of Chemistry, University of Massachusetts, Amherst, Mass. 01002.

yield of product. The preparation requires *ca.* 4 hours; no substantial modifications are necessary in order to scale the synthesis for larger quantities of product other than allowance for a slightly extended period of filtration and solvent evaporation.

Procedure

■ **Caution.** *Phosphorus(III) isocyanate is quite toxic and all operations should be carried out in an efficient hood.*

Silver isocyanate is prepared by the method of Neville and McGee.[6] The material obtained from this preparation should be thoroughly dried in a vacuum desiccator over P_4O_{10} for at least 2 days before use.

A 500-ml., three-necked flask is equipped with a Teflon blade stirrer, a reflux condenser, and an equilibrating dropping funnel. The top of the condenser is fitted with a gas inlet tube and provision made for maintaining an inert atmosphere throughout the preparation. The assembled apparatus should be carefully dried before use.

Into the flask is placed 100 g. (0.66 mole) of silver isocyanate and 100 ml. of dry benzene. A mixture of 30 g. (0.22 mole) of phosphorus(III) chloride and 20 ml. of dry benzene is placed in the dropping funnel. The entire system is then flushed once more with nitrogen. The mixture of phosphorus(III) chloride–benzene is allowed to flow in dropwise, with stirring, over a period of half an hour.*

The rate of addition should be regulated to avoid excessive generation of heat during the initial reaction. After addition is complete, the resulting mixture is refluxed for a period of $1\frac{1}{2}$ hours.

The cooled slurry is vacuum-filtered through a fritted tube assembly. Provision should be made for admitting nitrogen into the apparatus upon completion of the filtration. The sol-

---

* Rapid darkening of the silver isocyanate at this point indicates incomplete drying, and a significantly lower yield of product can be expected.

vent is removed at room temperature using a solvent stripper. The residual liquid is transferred, under nitrogen, to a semimicro distillation apparatus and the product distilled through a short column at 81°/20 mm. pressure to yield 18–21 g. of phosphorus(III) isocyanate (52–60% based on silver isocyanate). *Anal.** Calcd. for $P(NCO)_3$: C, 22.93%; N, 26.76%; P, 19.73%. Found: C, 22.90%; N, 26.70%; P, 19.65%.

Properties

Phosphorus(III) isocyanate boils at 169.3° (760 mm.), and melts at −2.0°.[1] Upon standing for 2–3 days polymerization occurs to yield a white material of m.p. 80–95°. The rate of polymerization is retarded by the presence of contaminants and serves as an indication of product purity. Distillation of the polymer melt results in nearly quantitative yields of the monomer.

The product has $n_D^{25°} = 1.525$ and $d_{26°} = 1.450$ g./cc. The specific conductivity is 1.89×10^{-5} Ω^{-1} at 25°C.[7] The infrared spectrum of phosphorus(III) isocyanate reveals fundamental absorptions at 316, 365, 388, 577, 603, 681, 1421, 2239, and 2293 cm.$^{-1}$. The band at 1421 cm.$^{-1}$ is the principal evidence cited for the isocyanate formulation in bonding to phosphorus.[8] Contamination by phosphoryl isocyanate can be detected by infrared absorption at 1282 cm.$^{-1}$, attributed to the P=O stretching frequency. The phosphine also exhibits ^{31}P n.m.r., absorption at −97.0 p.p.m. relative to 85% H_3PO_4.[9]

References

1. G. S. Forbes and H. H. Anderson, *J. Am. Chem. Soc.*, **62**, 761 (1940).
2. H. H. Anderson, *J. Am. Chem. Soc.*, **72**, 193 (1950).
3. L. H. Jenkins and D. S. Sears, U.S. Patent 2,873,171 (1959); *C. A.*, **53**, 18864f (1959).
4. Imperial Chemical Industries Ltd., British Patent 907,029 (1959); *C. A.*, **58**, 279d (1963).

* Analysis performed by checkers.

5. Imperial Chemical Industries Ltd., Belgian Patent 612,359 (1962); *C. A.*, **57**, 13411h (1962).
6. R. G. Neville and J. J. McGee, *Inorganic Syntheses*, **8**, 23 (1966).
7. E. Colton and L. S. Cyr, *J. Inorg. Nucl. Chem.*, **7**, 424 (1958).
8. F. A. Miller and W. K. Baer, *Spectrochim. Acta*, **18**, 1311 (1962).
9. E. Fluck, F. L. Goldman, and K. D. Rumpler, *Z. Anorg. Allgem. Chem.*, **338**, 52 (1965).

6. PHOSPHORAMIDIC ACID AND ITS SALTS

Submitted by R. C. SHERIDAN,* J. F. McCULLOUGH,* and Z. T. WAKEFIELD*
Checked by H. R. ALLCOCK† and E. J. WALSH†

The classical Stokes' method[1] for the preparation of phosphoramidic acid and its salts entails a three-step procedure with the use of organic intermediates. Ammonium hydrogen phosphoramidate, $NH_4HPO_3NH_2$, however, is prepared more conveniently and in higher yield by the simple reaction of phosphoryl chloride with aqueous ammonia. The acid salt is stable and nonhygroscopic, and it is readily converted to the free acid or to other salts of the acid.

A. AMMONIUM HYDROGEN PHOSPHORAMIDATE

$$POCl_3 + 5NH_3 + 2H_2O \rightarrow NH_4HPO_3NH_2 + 3NH_4Cl$$

Procedure

Reagent-grade phosphoryl chloride (18.3 ml., 0.2 mole) is added, dropwise and with vigorous stirring, for about 5 minutes to 300 ml. of an ice-cold 10% aqueous ammonia solution (1.5 moles NH_3). There is some fuming and evolution of heat, after

---

* Tennessee Valley Authority, Muscle Shoals, Ala. 35660.
† Pennsylvania State University, University Park, Pa. 16802.

which a clear solution is obtained. After stirring for about 15 minutes the solution is diluted with 1 l. of acetone,* whereupon two layers are formed.† The bottom layer is separated, neutralized to approximately pH 6 (Alkacid test paper) with 8 ml. of glacial acetic acid, and cooled to 5–10° to induce crystallization of the phosphoramidate. Further amounts of the salt are obtained by dilution of the filtrate with its own volume of ethanol. The product is filtered by suction, washed successively with alcohol and ether, and air-dried. Yield is 13.7 g. (0.12 mole), or 60% based on $POCl_3$. The ammonium hydrogen phosphoramidate is homogeneous and well-crystallized. *Anal.* Calcd. for $NH_4HPO_3NH_2$: N, 24.56; N in NH_4, 12.28; P, 27.14. Found: N, 24.5; N in NH_4, 12.4; P 27.0.

B. PHOSPHORAMIDIC ACID

$$NH_4HPO_3NH_2 + HClO_4 \rightarrow H_2PO_3NH_2 + NH_4ClO_4$$

Procedure

A stirred solution of 11.4 g. (0.1 mole) of the ammonium acid salt in 150 ml. of water is cooled to 0° in an ice bath, and 50 ml. of 25% $HClO_4$ (prepared by diluting 25 ml. of reagent 70% $HClO_4$ to 100 ml.) is added dropwise. The acid solution is diluted immediately with 1 l. of ethanol and allowed to stand in the ice bath for 30–60 minutes to induce crystallization.‡ The product is collected, washed successively with alcohol and ether, and air-dried. Yield is 7.5 g. (77%). *Anal.* Calcd. for $H_2PO_3NH_2$: N, 14.44; N in NH_4, 0.00; P, 31.93. Found: N, 14.4; N in NH_4, 1.3; P, 31.5.

---

* In the absence of an excess of ammonia some of the $NH_4HPO_3NH_2$ will precipitate at this stage.

† Paper chromatographic analysis of the liquid phases shows that all the phosphoramidate is in the aqueous (bottom) layer and the other amidophosphates, $NH_4PO_2(NH_2)_2$ and $PO(NH_2)_3$, are in the acetone layer.

‡ The checkers found that the longer time was required for high yield.

C. POTASSIUM HYDROGEN PHOSPHORAMIDATE

$$NH_4HPO_3NH_2 + 2KOH \rightarrow K_2PO_3NH_2 + 2H_2O + NH_3$$
$$K_2PO_3NH_2 + CH_3CO_2H \rightarrow KHPO_3NH_2 + CH_3CO_2K$$

Procedure

Ammonium hydrogen phosphoramidate (11.4 g., 0.1 mole) is dissolved in 20 ml. of 50% potassium hydroxide solution and warmed to 50–60° for 10 minutes to expel ammonia. The solution is cooled to 5–10°, neutralized to pH 6 (Alkacid test paper) with glacial acetic acid, and treated with 1 l. of ethanol to precipitate the potassium salt which is filtered with suction, washed successively with alcohol and ether, and air-dried. Yield is 8.1 g. (60%). *Anal.* Calcd. for $KHPO_3NH_2$: N, 10.37; P, 22.93. Found: N, 10.4; P, 22.7.

Properties

Ammonium hydrogen phosphoramidate forms colorless monoclinic crystals that appear to be orthorhombic. The powder x-ray diffraction[2] and infrared absorption[3] data have been reported. The salt is stable and nonhygroscopic, is moderately soluble in water, and melts at 233–234°.

Phosphoramidic acid forms colorless, prismatic crystals; the x-ray diffraction[2] and infrared absorption[4] data have been reported. The compound is easily soluble in water, but the aqueous solution hydrolyzes to ammonium dihydrogen phosphate. It is stable in sealed containers, but it converts on heating to an ammonium polyphosphate.[5] An unstable monohydrate of phosphoramidic acid has been described.[2]

Potassium hydrogen phosphoramidate is a white, crystalline compound that is moderately soluble in water and insoluble in alcohol. The x-ray diffraction[2] and infrared absorption[3] data have been published.

References

1. H. N. Stokes, *Am. Chem. J.*, **15**, 198 (1893).
2. A. H. Herzog and M. L. Nielsen, *Anal. Chem.*, **30**, 1490 (1958).
3. D. E. C. Corbridge and E. J. Lowe, *J. Chem. Soc.*, **1954**, 493.
4. E. Steger, *Z. Anorg. Allgem. Chem.*, **309**, 304 (1961).
5. M. Goehring and J. Sambeth, *Chem. Ber.*, **90**, 232 (1957).

7. DIPHENYL(TRIMETHYLSILYL)PHOSPHINE AND DIMETHYL(TRIMETHYLSILYL)PHOSPHINE

Submitted by R. GOLDSBERRY* and KIM COHN*
Checked by M. F. HAWTHORNE,† G. B. DUNKS,† and R. J. WILSON†

The first reported preparation of a silylphosphine in the literature involved heating silane and phosphine in a tube at 450° to give silylphosphine (H_3SiPH_2).[1] A general procedure for making unsubstituted silicon–phosphorus compounds or ones that possess more than one silicon–phosphorus bond involves the interaction of *n*-butyl lithium with phosphine which gives a mixture of lithium phosphides. The latter are then allowed to react with the appropriate chlorosilane to give the mono-, di-, and trisubstituted silylphosphines.[2] For the preparation of silylphosphines with phenyl or alkyl substituents on the phosphorus, a better method involves the reaction of sodium or lithium metals with the tetraphenyl or tetraalkyl diphosphines. Diphenyl(trimethylsilyl)phosphine has been prepared in yields above 60% by the reaction of chlorotrimethylsilane with sodium diphenylphosphide in *n*-butyl ether.[3] The sodium diphenylphosphide is prepared from commercially available diphenylphosphinous chloride. Although the initially formed product is the tetraphenyldiphosphine, the phosphorus–phos-

* Department of Chemistry, Michigan State University, East Lansing, Mich. 48823.
† University of California, Riverside, Calif. 94802.

phorus bond is cleaved by the action of excess sodium to give the sodium salt.

Diethyl(trimethylsilyl)phosphine has been prepared by the reaction of lithium diethylphosphide with chlorotrimethylsilane in ether solution.[4] The lithium diethylphosphide may be prepared by the reaction of an ether solution of phenyllithium with diethylphosphine.[5] However, the dialkylphosphines are most conveniently prepared by the reduction of the corresponding tetraalkyldiphosphine disulfides with lithium tetrahydroaluminate in ether.[6,7] An alternative method for the preparation of dimethyl(trimethylsilyl)phosphine which eliminates the handling of the volatile dimethylphosphine involves the preparation of lithium dimethylphosphide from tetramethyldiphosphine. The latter is prepared by the reduction of tetramethyldiphosphine disulfide[8] with tributylphosphine.[9] The reaction of chlorotrimethylsilane with lithium dimethylphosphide is most conveniently carried out in a vacuum system without solvent at $-78°$.

■ *Caution. The silylphosphines and many of the intermediates described in the following preparations are very sensitive to oxygen and moisture. All of the phosphines are extremely toxic, and the alkyl phosphines are spontaneously flammable in air. All operations involving these materials should be carried out in an inert atmosphere and in a good hood.*

A. DIPHENYL(TRIMETHYLSILYL)PHOSPHINE

$$2(C_6H_5)_2PCl + 2Na \xrightarrow[\text{ether}]{\textit{n}\text{-butyl}} 2NaCl + (C_6H_5)_2P—P(C_6H_5)_2$$

$$(C_6H_5)_2P—P(C_6H_5)_2 + 2Na \xrightarrow[\text{ether}]{\textit{n}\text{-butyl}} 2(C_6H_5)_2PNa$$

$$(CH_3)_3SiCl + (C_6H_5)_2PNa \xrightarrow[\text{ether}]{\textit{n}\text{-butyl}} (CH_3)_3SiP(C_6H_5)_2 + 2NaCl$$

Procedure

A 250-ml., pressure-equalized dropping funnel and nitrogen inlet tube are inserted through one of the sidearms of a carefully dried 500-ml., three-necked, round-bottomed flask. A stirring rod is inserted through the main mouth of the flask, and the other sidearm is equipped with an Allihn condenser fitted with a drying tube. The flask is flushed with a slow stream of nitrogen, and a suspension of 15.22 g. (0.65 moles) of sodium sand[10] in 300 ml. of *n*-butyl ether (previously dried over sodium and benzophenone) is heated so that the solvent is refluxing and is stirred by means of a mechanical stirrer. Diphenylphosphinous chloride (33 g., 0.15 mole, 26.8 ml.) with 75 ml. of *n*-butyl ether (dried over sodium and benzophenone) is added drop by drop over a one-hour period. The addition of the first portions of diphenylphosphinous chloride results in a bright yellow solution. After the addition of all of the diphenylphosphinous chloride the mixture becomes amber and brown solids precipitate. This mixture is allowed to reflux for a total of 4 hours. After this time the suspension of sodium diphenylphosphide and sodium chloride is transferred under a stream of nitrogen to a 1-l. three-necked round-bottomed flask. The excess sodium and the brown precipitate remain behind. ■ *Caution. These residues will inflame spontaneously when exposed to air and should be disposed of out-of-doors. These materials should be handled only where fire extinguishers, a fire blanket, and safety showers are available. Safety glasses should be supplemented with a face shield. Asbestos gloves and a nonflammable coat should also be worn. The flask containing the residues is cooled to room temperature and thoroughly flushed with nitrogen. The excess sodium and other residues are decomposed by careful drop-by-drop addition of a dilute solution of water in dioxane or tetrahydrofuran to the stirred and cooled reaction mixture until hydrogen evolution stops. The mixture is then diluted with 500 ml. of water, and the residual traces of the phosphine are destroyed by addition of sodium hypochlorite solution. (A commercial bleaching agent such as Clorox is satisfactory.) As an added precaution, the decomposition should be done behind a safety shield in the event that explosion of the evolved hydrogen should occur.*

Using equipment similar to that employed in the preparation of sodium diphenylphosphide, chlorotrimethylsilane (49.5 g., 0.45 mole, 59 ml.) dissolved in 100 ml. of dry *n*-butyl ether is added drop by drop to the refluxing, stirred suspension of sodium diphenylphosphide over a 2-hour period. After refluxing for an additional hour, the mixture is filtered and washed under nitrogen by means of a filtering stick similar to one previously described.[11] The solvent is removed by distillation at atmospheric pressure, and the residual oil is fractionally distilled at 1 mm. pressure in a nitrogen atmosphere. In a typical experiment a 24.0-g. (62%) fraction is collected as product (b.p. 126–127°/1 mm., n_D^{25} 1.600).

Properties

Diphenyl(trimethylsilyl)phosphine is a colorless liquid that yellows upon standing. Like diphenylphosphine, it has a very bad odor and is reactive toward water and oxygen. The ¹H n.m.r. spectrum shows a doublet ($J_{P-H} = 4.9$ Hz.) which is assigned to methyl protons on silicon at $\delta = -0.14$ p.p.m. (TMS) and a complicated multiplet from $\delta = -7.0$ to -8.6 p.p.m. which can be assigned to the phenyl protons.

B. DIMETHYL(TRIMETHYLSILYL)PHOSPHINE

$$\overset{\text{S}}{\underset{\uparrow}{\text{Me}_2\text{P}}}-\overset{\text{S}}{\underset{\uparrow}{\text{PMe}_2}} + 2(n\text{-Bu})_3\text{P} \rightarrow 2(n\text{-Bu})_3\overset{\text{S}}{\underset{\uparrow}{\text{P}}} + \text{Me}_2\text{P}-\text{PMe}_2$$

$$\text{Me}_2\text{P}-\text{PMe}_2 + 2\text{Li} \xrightarrow{\text{Et}_2\text{O}} 2\text{Me}_2\text{PLi}$$

$$\text{Me}_2\text{PLi} + \text{Me}_3\text{SiCl} \xrightarrow{\text{neat}} \text{Me}_3\text{SiPMe}_2 + \text{LiCl}$$

Procedure

■ *Caution.* *All residues (approximately 200 ml. of material remaining after the initial distillation, the vacuum-line trap into which the ethyl ether is condensed, the solid residues which remain behind after the product is removed from the reaction flask, and particularly the vacuum-line trap which contained the ethyl ether and unreacted tetramethyldiphosphine) may inflame spontaneously. The precautions described in the preparation of diphenyl(trimethylsilyl)phosphine should be observed while handling these residues. If a dilute solution of water with a nitrogen atmosphere cannot be added to these residues (e.g., the vacuum-line trap), it is suggested that the residues be distilled in vacuo into a reaction flask equipped with a stopcock and a standard-taper joint which can be removed from the vacuum line. This flask can then be removed from the vacuum line and handled in the manner previously described.*

It should also be remembered that the spontaneously flammable material may be trapped in the stationary center tube of a vacuum-line trap; therefore the use of U-type traps is suggested, so that the entire trap and its contents can be removed. The residues should be kept under an inert atmosphere until they can be destroyed.

A carefully dried, 500-ml., three-necked, round-bottomed flask equipped with a thermometer, nitrogen inlet tube, and magnetic stirrer is fitted to an insulated Vigreux column 6 in. long. A standard distilling head with a thermometer and a 200-mm. Liebig condenser are connected to the column. The receiving flask is a 250-ml., two-necked, round-bottomed flask which is equipped with a nitrogen inlet tube and a vacuum adapter. The end of the vacuum adapter has a drying tube. The receiving flask is surrounded by a Dewar containing a slush bath of Dry Ice and isopropyl alcohol. Commercial tributylphosphine (81.52 g., 0.40 moles) and tetramethyldiphosphine disulfide (37.4 g., 0.20 moles) are placed in the distilling flask, and the system is thoroughly flushed with nitrogen. It is very important to maintain a positive pressure of nitrogen in the system throughout the distillation. The mixture is stirred for 5 minutes and slowly heated until the temperature of the distilling flask is about 250°. As the temperature rises, the mixture becomes homogeneous, and the product, tetramethyldiphosphine, which boils at 100–110° is collected. Yield is 19.7 g. (81%, 0.16 mole). The receiving flask is flushed with nitrogen through the inlet tube on the receiving flask and stoppered. In a glove bag filled with nitrogen 2.2 g. (0.32 mole) of lithium chips and 100 ml. of diethyl ether are added to the tetramethyldiphosphine. The flask is removed from the glove bag and fitted with a condenser containing a drying tube while maintaining a slow flow of nitrogen throughout the system. The mixture is stirred and heated for 8 hours during which time a white solid is formed. The receiving flask is then fitted with a vacuum-stopcock adapter and attached to a vacuum system. The ether is removed *in vacuo* and the flask is surrounded by a Dewar of liquid nitrogen (−196°). Commercial chlorotrimethylsilane, b.p. 57° (40 g., 0.37 mole, 47 ml.), which has been previously distilled, is then distilled *in vacuo* at −196° into the flask, and the system is then allowed to warm slowly to −160° (isopentane slush) and then again allowed to warm slowly to −63° (chloroform slush). During this time, a slow reaction takes place between the lithium dimethylphosphide and the chlorotrimethylsilane. Periodically the stopcock on the reaction flask is closed and the flask is removed from the vacuum system and shaken to ensure complete mixing of the reactants. After 3 hours the system is slowly warmed to −45° (chlorobenzene slush), and all volatile products are distilled *in vacuo* to a trap held at −196°. The impure dimethyl(trimethylsilyl)phosphine which remains in the reaction flask with the solid lithium chloride is permitted to warm to room temperature and distilled *in vacuo* into a 100-ml. flask held at −196°. Finally, the dimethyl-(trimethylsilyl)phosphine trapped in the 100-ml. flask is distilled at 20 mm. pressure in a nitrogen atmosphere to yield 25.0 g. (58%) of product, b.p. 33–34°/20 mm. The product can also be purified by careful fractional distillation on the vacuum line from 0 to −50° (*n*-hexanol slush) to −196°. The product collects at −50°.

Properties

Dimethyl(trimethylsilyl)phosphine is a spontaneously flammable, colorless liquid with an extrapolated normal boiling point of 130°. It has an obnoxious odor and is very reactive to moisture and air. Since the dimethyl(trimethylsilyl)phosphine hydrolyzes to dimethylphosphine, it should always be handled in a good, well-ventilated hood. The ^{1}H n.m.r. spectrum shows a doublet (J_{P-H} − 4.5 Hz.) at δ = −0.13 p.p.m. which is assigned to the methyl protons on silicon and a doublet (J_{P-H} = 2.5 Hz.) which is assigned to methyl protons on phosphorus at δ = −1.0 p.p.m.

References

1. G. Fritz, *Z. Naturforsch.*, **8b**, 776 (1953).
2. G. W. Parshall and R. V. Lindsey, Jr., *J. Am. Chem. Soc.*, **81**, 6273 (1959).
3. W. Kuchen and H. Bachwald, *Chem. Ber.*, **92**, 227 (1959).
4. G. Fritz and G. Poppenburg, *Angew. Chem.*, **72**, 208 (1960).
5. K. Issleib and A. Tzschach, *Chem. Ber.*, **92**, 1125 (1959).
6. K. Issleib and A. Tzschach, *ibid.*, **92**, 704 (1959).
7. G. W. Parshall, *Inorganic Syntheses*, **11**, 157 (1968).
8. G. W. Parshall, *Org. Syn.*, **45**, 102 (1965).
9. L. Maier, *J. Inorg. Nucl. Chem.*, **24**, 275 (1962).
10. W. Jolly, "Inorganic Syntheses," p. 161, Prentice-Hall, Inc., Englewood Cliffs, N.J., 1960.
11. D. Holah and J. P. Fackler, Jr., *Inorganic Syntheses*, **10**, 29 (1961).

8. EXCHANGE REACTIONS FOR THE SYNTHESES OF PHENYLBORON CHLORIDES AND TETRAFLUOROMETHYLPHOSPHORANE

Submitted by P. M. TREICHEL,* J. BENEDICT,* AND RUTH GOODRICH HAINES*
Checked by B. GASSENHEIMER† and T. WARTIK†

Among possible methods available to accomplish the synthesis of an alkyl- or aryl-metal or -metalloid compound, organo-

* University of Wisconsin, Madison, Wis. 53706.
† Pennsylvania State University, University Park, Pa. 16802.

metallic exchange reactions are the best known. These reactions, of the form $R—M + M'—X \leftrightarrows R—M' + M—X$, will in general proceed to the products on the right if the metal M is more electropositive than M'.[1] Commonly, the reactive organometallic compounds of electropositive metals such as lithium or magnesium are used as alkylating agents in these reactions. Less reactive organometallic compounds also can serve in this capacity; the best-known examples probably are diorganomercury and tetraorganotin compounds. Organotin reagents are particularly appropriate in syntheses of organoboron halides from boron trihalides. Such reactions are described here.

There are several advantages to the use of tetraorganotin reagents in the syntheses of organoboron halides. Foremost is the fact that partial alkylation to mono- or dialkyl-boron halides (RBX_2 or R_2BX) can be controlled by choice of reaction stoichiometry. In contrast, reactions of the more active organometallic reagents are best employed in syntheses of fully substituted R_3B compounds, but are difficult to adapt in RBX_2 and R_2BX preparations. Furthermore the use of organolithium or Grignard reagents is often convenient only in coordinating solvents which can be of some disadvantage since the boron compounds can coordinate strongly to the solvent also, making isolation of the pure product more difficult. It is noted that many tetraorganotin compounds are commercially available; they are easily stored, can be handled without excessive concern for oxygen or moisture sensitivity, and are easily measured out by weight or volume.

Exchange reactions between tetraalkyltin compounds and boron trichloride were first reported by Stone and coworkers. After a preliminary communication in 1959 which implied the feasibility of such reactions in the synthesis of organoboron halides,[2] two papers were published in 1960 describing in more detail preparations of $CH_2{=}CHBCl_2$, $(CH_2{=}CH)_2BCl$, $(CH_2{=}CH)_3B$,[3] and $CF_2{=}CFBCl_2$, $(CF_2{=}CF)_2BCl$, and $(CF_2{=}CF)_3B$[4] from appropriate organotin compounds. At about the same time two papers appeared independently describing the preparation of dichloro(phenyl)borane from tetraphenyl-stannane and boron trichloride.[5,6] In 1964, a more detailed study of the tetraalkylstannane-boron trihalide (bromide and chloride) reactions was published.[7]

For reactions of tetraalkyltin compounds without solvent two of the four alkyl groups can be transferred to boron. The reaction stoichiometry can thus be assigned to give maximum conversion to the desired product. With tin compound to boron trihalide in a 1:2 molar ratio, the monoalkyl boron compounds RBX_2 predominate, and with a 1:1 ratio the dialkylboron compounds, R_2BX, are the main products. For higher ratios of $R_4Sn:BX_3(3:2)$[7] some of the trialkylboron compound is formed but the conversion does not seem to be good:

$$R_4Sn + 2BX_3 \rightarrow R_2SnX_2 + 2RBX_2$$
$$R_4Sn + BX_3 \rightarrow R_2SnX_2 + R_2BX$$

The reaction of tetraphenylstannane and boron trichloride in refluxing benzene is reported to lead to cleavage of only two of the phenyl groups from tin,[6] whereas in refluxing methylene chloride three groups are cleaved and in refluxing carbon tetra-chloride all four phenyl groups can be transferred to boron.[5] Using boron tribromide, without solvent, only two phenyl groups will be transferred to boron. Though reported reactions describe only dichloro(phenyl)borane as a product of these reactions, one can adjust the reaction stoichiometry to allow preparation of chlorodiphenylborane as well. The temperature at which such reactions are run is not clearly stated in the references, but our observations suggest that temperatures of about 80° are appropriate. At lower temperatures reactions are slow, whereas higher temperatures lead to formation of hydrocarbons.[7]

It is of interest that reactions of tetraorganotin compounds and boron trifluoride differ somewhat from the reactions of the other boron trihalides. The reactions of boron trifluoride proceed in a less facile manner, and yield monosubstituted organo-difluoroboranes only. Thus Stone and coworkers reported the syntheses of $CH_2{=}CHBF_2$, $n\text{-}C_3H_7BF_2$, $CH_2{=}CHCH_2BF_2$,[2] CH_3BF_2, and $C_2H_5BF_2$[8] from reactions of boron trifluoride and the corresponding tetraorganotin compound. Reactions with (perfluorovinyl)tin compounds were reported to give $CF_2{=}$ $CFBF_2$.[9] Burg and Spielman subsequently studied the reaction of boron trifluoride and tetramethylstannane in more detail.[10] The reaction of tetraphenylstannane and boron trifluoride to give difluoro(phenyl)borane has also been reported.[11]

A. DICHLORO(PHENYL)BORANE, $C_6H_5BCl_2$

$$(C_6H_5)_4Sn + 2BCl_3 \rightarrow (C_6H_5)_2SnCl_2 + 2C_6H_5BCl_2$$

Tetraphenylstannane (68.0 g., 0.17 mole) and 700 ml. dry benzene were placed in a 2-l., three-neck flask fitted with stirrer, nitrogen inlet, and Dry-Ice condenser. The mixture was cooled in an ice bath to *ca.* 5°. Then 30 ml. of liquid boron trichloride (*ca.* 42 g., 0.36 mole) was rapidly poured into the flask through one inlet. (See notes 1 and 2.) The mixture was warmed to room temperature with stirring, then brought to reflux temperature and held there for $3\frac{1}{2}$ hours. Then the condenser was replaced by a distillation head and collection assembly and the stirrer replaced by a thermometer extending to the bottom of the flask. Most of the solvent was removed by distillation (until a pot temperature of *ca.* 140° was reached). The remaining liquid was then transferred to a smaller flask under a nitrogen stream and the distillation continued, giving a fraction collected in the range 150–200°. This was redistilled at reduced pressure to give 57.2 g. clear liquid dichloro(phenyl)borane, b.p. 38–39°/1 mm. (literature,[5] b.p. 47°/5 m.m., 67–70°/7 m.m., 82–86°/30 mm.). Yield is 56%.

Properties

Dichloro(phenyl)borane is a clear liquid. It is sensitive to water, hydrolyzing to dihydroxy(phenyl)borane $C_6H_5B(OH)_2$.

It is most conveniently stored in glass ampuls. Several other preparations for this compound are known including reactions of dihydroxy(phenyl)borane and boron trichloride[12, 13] or phosphorus pentachloride,[14] 2,4,6-triphenylboroxin, $[C_6H_5BO]_3$, with phosphorus pentachloride,[14] phenyl mercury chloride with boron trichloride,[15] and benzene with boron trichloride.[16]

B. CHLORODIPHENYLBORANE, $(C_6H_5)_2BCl$

$$(C_6H_5)_4Sn + BCl_3 \rightarrow (C_6H_5)_2BCl + (C_6H_5)_2SnCl_2$$

Tetraphenylstannane (70 g., 0.16 mole) was added to a three-necked flask containing 300 ml. dry benzene. The flask had a nitrogen inlet, Dry Ice condenser, and magnetic stirrer. A very slow nitrogen stream was continued through the reaction. After cooling the mixture to 10°, *ca.* 13 ml. liquid boron trichloride (*ca.* 18 g., 0.15 mole) was quickly poured into the flask. (See notes 1 and 2.) The mixture was warmed slowly to gentle reflux with stirring. After 3 hours the Dry Ice condenser was replaced by a water condenser and the mixture refluxed for 48 hours.

At this time most of the benzene was distilled off, and the remaining material transferred under nitrogen to another flask for vacuum distillation. After distillation of the remaining benzene, a fraction was collected, which boiled in the range 110–118°/3 mm. This was redistilled to give 21.0 g. of chlorodiphenylborane, b.p. 112–113°/4 mm. (literature,[13] b.p. 98°/0.1 mm., m.p. 21.5–22°). Yield is 75%.

Properties

Chlorodiphenylborane is a clear liquid which slowly hydrolyzes. It is best stored for prolonged periods of time in glass ampuls. It has also been prepared by the reaction of oxybis-(diphenylborane) and boron trichloride,[13] the former reagent prepared in a several-step reaction sequence from tetraphenyl-borate anion.

Notes

1. An approximate weight of boron trichloride is sufficient for most purposes, and can best be obtained as follows: Obtain a dry tube of sufficient size calibrated by volume (often a large centrifuge tube will suffice). Place a two-holed stopper in the top with short inlet and outlet tubes, and after chilling the tube at $-78°$ (Dry Ice–trichloroethylene), pass boron trichloride slowly into it (in a hood). After collecting approximately the right amount of boron trichloride allow the tube to warm slowly to $0°$, with a very slow stream of boron trichloride to prevent moisture from condensing in the tube. Measure the boron trichloride volume in the tube (density at $11° = 1.40$).

2. Boron tribromide is not appropriate in these experiments as it will not be easily freed from phenyltin bromide which has the same volatility.

C. TETRAFLUOROMETHYLPHOSPHORANE, CH_3PF_4

$$(CH_3)_4Sn + PF_5 \rightarrow CH_3PF_4 + (CH_3)_3SnF[+(CH_3)_3SnPF_6]$$

Alkyltetrafluorophosphoranes have been prepared by fluorination of complexes $RPCl_3^+AlCl_4^-$ using hydrogen fluoride, arsenic, or antimony trifluorides or alkali-metal fluorides,[16,17] by oxidation and fluorination of alkyldichlorophosphines with arsenic or antimony trifluorides and pentafluorides,[18,19] by chlorination of alkyldichlorophosphines followed by fluorination,[17,20] and by fluorination, using sulfur tetrafluoride, of phosphonic acids, phosphinic fluorides, or phosphine oxides.[20,21] The reaction described here[22] provides a method for the small-scale synthesis of tetrafluoromethylphosphorane and other lower members of the RPF_4 series from commercially available starting materials. The method utilizes a small vacuum system of standard construction, and it avoids the use of highly reactive intermediates or fluorinating reagents necessary for other methods of synthesis.

The analogous reaction of tetraphenylstannane and phosphorus pentafluoride in a steel bomb reactor at 135° is reported to give tetrafluoro(phenyl)phosphorane.[11]

Procedure

All manipulations of volatile reactants and products are accomplished with a standard Pyrex vacuum system. Kel-F stopcock grease is inert to the species to be handled and is recommended though it is somewhat thin. Apiezon(N) tends to darken rapidly on contact with phosphorus(V) fluorides, but can be used successfully for short periods of time and for manipulations necessary for rough transfer and purification.

Tetraalkylstannanes and phosphorus pentafluoride are commercially available. Commercial phosphorus pentafluoride contains a small amount of phosphoryl fluoride, but its presence does not hinder the exchange reaction or cause any subsequent difficulties in workup of the products. It may be removed if desired, however, by carefully fractionating the commercial sample through −112° (CS_2 slush) and −196° traps. The former trap will retain the phosphoryl fluoride, whereas the pentafluoride will pass this trap and collect at −196°. The vapor pressure of phosphorus pentafluoride at −96° (toluene slush) is 335 mm. Hg; the purity of the sample may be checked by comparison of its vapor pressure against this value.

The amounts of phosphorus pentafluoride to be reacted can be measured in the gas phase in a calibrated volume. Aliquots of tetramethylstannane are best measured from a preweighed ampul from which this reactant can be distilled into the vacuum system.

Phosphorus pentafluoride (32.0 mmoles) and tetramethylstannane (3.961 g., 22.2 mmoles) are transferred from the line into a flame-dried, evacuated, 500-cc. reaction bulb. The bulb is then sealed from the line, warmed to room temperature, and then heated at 60° for about 6 hours. Alternatively the reaction can be run at room temperature for about 7 days. During this time a white solid forms inside the bulb. The bulb is then reattached to the vacuum system and opened. The volatile contents are fractionated through a series of traps maintained at −64°. (chloroform slush), −96° (toluene slush), and −196°, with the contents of the −64° and −96° trap then being recycled several times. Gas-phase infrared spectra at pressures of 20–30 mm. are used to identify fractions and to help judge the extent of fractionation.[7] The −196° trap collects 4.6 mmoles of phosphorus pentafluoride. The −96° trap collects 19.2 mmoles of the product tetrafluoromethylphosphorane, and the −64° trap collects 0.503 g. (2.8 mmoles) tetramethylstannane. This overall conversion to tetrafluoromethylphosphorane is 87%.

The white solid remaining in the bulb fumes in moist air and is partially hydrolyzed to give an acid solution (HF). It can be identified as a mixture of fluorotrimethylstannane and trimethyltin hexafluorophosphate.[22] The latter substance cannot be isolated as a pure sample.

A slight modification of the above description is necessary if this reaction is to be used for the synthesis of other alkylfluorophosphoranes ($CH_3CH_2PF_4$, $CH_2{=}CHPF_4$, $n\text{-}CH_3CH_2CH_2PF_4$). Because all other tetraalkylstannanes are not transferable in a vacuum system at 25°, they can be syringed into the reaction bulb before attachment to the vacuum lines. Care to use dry samples and equipment must be observed, however, since the presence of even small amounts of water will lead to formation of phosphoryl fluoride and silicon tetrafluoride.

Properties[23]

Tetrafluoromethylphosphorane is a volatile liquid with a reported boiling point of 10–12.5°[17, 26] and melting point of

$-50°.[24]$ It has vapor pressures of 139 mm. Hg at $-24°$, 9.6 mm. Hg at $-63.5°$. The vapor-pressure curve of the liquid follows the equation $\log P_{mm.\ Hg} = -1448.9/T + 7.9675$. It is thermally stable and can be handled in glass systems, but is eventually hydrolyzed to methylphosphonic acid by any traces of moisture. (CH_3POF_2 is an intermediate in the hydrolysis.) Tetrafluoromethylphosphorane behaves as a weak Lewis acid, forming complexes with strong bases such as amines and pyridine. It is also reported to function as a very weak Lewis base.[25] The gas-phase infrared spectrum of the phosphorane has been discussed in detail.[26] (The spectra of the ethyl, vinyl, and *n*-butyl analogs are also reported.[22]) In this synthesis the major contaminant in the product is tetramethylstannane which can be identified in the gas-phase infrared spectrum by the appearance of the strong band at 764 cm.$^{-1}$.[27] The presence of the stannane can also be ascertained by the molecular-weight determinations of the phosphorane via vapor density measurements.

References

1. P. L. Pauson, "Organometallic Chemistry," pp. 39–44, St. Martin's Press, Inc., New York, 1967; G. E. Coates, M. L. H. Green, and K. Wade, "Organometallic Compounds," 3d ed., Vol. 1, pp. 179–181, Methuen & Co., Ltd., London (1967).
2. F. E. Brinckman and F. G. A. Stone, *Chem. Ind.* (London), **1959**, 254.
3. F. E. Brinckman and F. G. A. Stone, *J. Am. Chem. Soc.*, **82**, 6218 (1960).
4. S. L. Stafford and F. G. A. Stone, *ibid.*, **82**, 6238 (1960).
5. J. E. Burch, W. Gerrard, M. Howarth, and E. F. Mooney, *J. Chem. Soc.*, **1960**, 4916.
6. K. Niedenzu and J. W. Dawson, *J. Am. Chem. Soc.*, **82**, 4223 (1964).
7. W. Gerrard, E. F. Mooney, and R. G. Rees, *J. Chem. Soc.*, **1964**, 740.
8. T. D. Coyle and F. G. A. Stone, *J. Am. Chem. Soc.*, **82**, 6223 (1960).
9. S. L. Stafford and F. G. A. Stone, *ibid.*, **82**, 6238 (1960); preliminary communication, H. D. Kaesz, S. L. Stafford, and F. G. A. Stone, *ibid.*, **81**, 6336 (1959).
10. A. B. Burg and J. R. Spielman, *ibid.*, **83**, 2667 (1961).
11. D. W. A. Sharp and J. M. Winfield, *J. Chem. Soc.*, **1965**, 2278.
12. P. A. McCusker and H. S. Makowski, *J. Am. Chem. Soc.*, **79**, 5185 (1957).
13. E. W. Abel, S. H. Dandegaonker, W. Gerrard, and M. F. Lappert, *J. Chem. Soc.*, **1956**, 4697.
14. S. H. Dandegaonker, W. Gerrard, and M. F. Lappert, *ibid.*, **1957**, 2895.

15. E. L. Muetterties, *J. Am. Chem. Soc.*, **82**, 4163 (1960).
16. British Patent 734,187 (1955); *C.A.*, **50**, P7123a (1956); U.S. Patent 2,853,515 (1958); *C.A.*, **53**, 7988g (1959).
17. I. P. Komkov, S. Z. Ivin, K. W. Karavanov, and L. E. Smirnov, *Zh. Obshch. Khim.*, **32**, 301 (1962).
18. U.S. Patent 2,904,588 (1959); *C.A.*, **54**, P2254d (1959).
19. R. Schmutzler, *Inorg. Chem.*, **3**, 410 (1964).
20. L. M. Yagupol'skii and Zh. M. Ivanova, *Zh. Obshch. Khim.*, **29**, 3766 (1959); *ibid.*, **30**, 4026 (1960).
21. W. C. Smith, *J. Am. Chem. Soc.*, **82**, 6176 (1960).
22. P. M. Treichel and R. A. Goodrich, *Inorg. Chem.*, **4**, 1424 (1965).
23. R. Schmutzler, *Chem. Ind.*, **1962**, 1868.
24. E. L. Muetterties and W. Mahler, *Inorg. Chem.*, **4**, 119 (1965).
25. A. J. Downs and R. Schmutzler, *Spectrochim. Acta.*, **21**, 1927 (1965).
26. W. F. Edgell and C. H. Ward, *J. Mol. Spectry.*, **8**, 343 (1962).

9. 2,4,6-TRICHLOROBORAZINE

$$BCl_3 + CH_3CN \rightarrow CH_3CN:BCl_3$$
$$3CH_3CN:BCl_3 + 3NH_4Cl \rightarrow B_3Cl_3N_3H_3 + 9HCl + 3CH_3CN$$

Submitted by D. T. HAWORTH*
Checked by W. G. PEET† and E. L. MUETTERTIES†

This boron–nitrogen heterocycle is a convenient route to the synthesis of borazine and *B*-substituted borazines.[1-6] The procedure described here is a modification of the reaction of boron trichloride with ammonium chloride.[7-9]

Procedure

A 1-l. three-necked flask is equipped with a stirrer, a reflux condenser, and a cold finger (Dewar condenser). A second cold finger is placed at the exit condenser, and it is protected by

* Department of Chemistry, Marquette University, Milwaukee, Wis. 53233.
† Central Research Department, Experimental Station, E. I. du Pont de Nemours & Company, Wilmington, Del. 19898.

a calcium chloride drying tube. A piece of rubber tubing equipped with a pinch clamp connects the cold finger inserted into the three-necked flask to a laboratory-tank (bottle) of boron trichloride. This size tank of boron trichloride is convenient, since it can be easily weighed on a trip scale or single-pan balance. The apparatus is dried before use by passage of a stream of nitrogen into the setup.

Reagent-grade ammonium chloride is dried at 110°. Chlorobenzene is dried over anhydrous sodium sulfate.

The reaction vessel is charged with 250–300 ml. of chlorobenzene and 16.4 g. (0.40 mole) of acetonitrile. The entrance and exit cold fingers are filled with a Dry Ice–acetone mixture. The preweighed tank of boron trichloride is opened, and the boron trichloride is allowed to enter the reaction vessel dropwise from the entrance cold finger. With moderate stirring, a white insoluble adduct of acetonitrile–boron trichloride adduct is formed. The boron trichloride tank is periodically weighed, and addition of the boron trichloride is continued until 50 g. (0.42 mole) has been added. This usually requires about 2 hours. After addition, the entrance cold finger is removed and 21.5 g. (0.4 mole) of ammonium chloride is added to the reaction vessel. The cold finger is replaced by a ground-glass stopper. The reaction mixture is refluxed over a period of 5 hours until the evolution of hydrogen chloride has almost ceased. The solution under nitrogen is transferred to a one-necked, 500-ml. flask, and the solvent is removed by vacuum distillation on a rotary evaporator. The remaining solids are sublimed at 45° in vacuum to a trap cooled to −10°. The yield of 2,4,6-trichloroborazine is 17–18 g. (*ca.* 70%, m.p. 83–84°).

Properties

2,4,6-Trichloroborazine prepared as above forms white crystalline needles. The product is very sensitive to moisture and should be removed from the sublimator in a dry-box and stored in a nitrogen-filled desiccator in a refrigerator. If the product is to be stored over a long period, resublimation will be necessary to obtain the pure compound. The infrared spectrum contains major absorption bands at 3442, 1445, 1031, 744, and 706 cm.$^{-1}$. Its ^{1}H n.m.r. spectrum has a broad absorption at -5.2 p.p.m. (TMS standard).

References

1. K. Niedenzu and J. W. Dawson, *Inorganic Syntheses*, **10**, 142 (1967).
2. G. E. Ryschkewitsch, J. J. Harris, and H. H. Sisler, *J. Am. Chem. Soc.*, **80**, 4515 (1958).
3. H. J. Becher and S. Frick, *Z. Anorg. Allgem. Chem.*, **295**, 83 (1958).
4. A. Groszos and S. F. Stafiej, *J. Am. Chem. Soc.*, **80**, 1357 (1958).
5. J. Pellon, W. G. Deichert, and W. M. Thomas, *J. Polymer. Sci.*, **55**, 153 (1961).
6. D. T. Haworth and L. F. Hohnstedt, *J. Am. Chem. Soc.*, **82**, 3860 (1960).
7. C. A. Brown and A. W. Laubengayer, *ibid.*, **77**, 3699 (1955).
8. E. F. Rothgery and L. F. Hohnstedt, *Inorg. Chem.*, **6**, 1065 (1967).
9. K. Niedenzu and J. W. Dawson, *Inorganic Syntheses*, **10**, 139 (1967).

10. 2,4,6-TRICHLORO-1,3,5-TRIMETHYLBORAZINE

$$3BCl_3 + 3CH_3NH_3Cl \rightarrow B_3Cl_3N_3(CH_3)_3 + 9HCl$$

Submitted by D. T. HAWORTH*
Checked by W. G. PEET† and E. L. MUETTERTIES†

The literature contains many references to the *N*-alkylated *B*-trichloroborazines, (BClNR)$_3$, as a starting material in the synthesis of boron-substituted borazines.[1] The procedure described is a modification of syntheses originally used in the preparation of 2,4,6-trichloroborazine.[2]

* Department of Chemistry, Marquette University, Milwaukee, Wis. 53233.
† Central Research Dept., Experimental Station, E. I. du Pont de Nemours & Company, Wilmington, Del. 19898.

Procedure

A 500-ml., three-necked flask is equipped with a mechanical stirrer, water-cooled condenser topped with a cold finger having a calcium-chloride exit tube, and an entrance cold finger connected by rubber tubing to a preweighed laboratory tank of boron trichloride. Under a blanket of nitrogen, the reaction flask is charged with a slurry of 25 g. (0.37 mole) of methylamine hydrochloride in 250–300 ml. of anhydrous sodium-sulfate-dried chlorobenzene. The mixture is stirred and heated to a gentle reflux. The cold fingers are charged with Dry Ice, and the boron trichloride is allowed to drip into the hot mixture at a rate of about one drop per 2 seconds. Periodically, the boron trichloride tank is removed and weighed; the rubber tubing connecting the entrance condenser is closed by a pinch clamp or by insertion of a glass rod into the tubing. Addition of the boron trichloride is continued until 55 g. (0.47 mole) has been added. The entrance condenser is removed and replaced by a ground-glass stopper. Heating is continued until the evolution of hydrogen chloride has almost ceased (about 15–18 hours). Excess boron trichloride is required since some of the boron trichloride passes the exit cold finger. The hot reaction mixture is usually light brown in color and it is generally clear. If some solid remains, the hot solution is filtered to a one-necked, 500-ml. flask and the chlorobenzene is removed by vacuum distillation on a rotary evaporator. The pasty solid is purified by vacuum sublimation at 60° into a trap at 0°. The product is removed in a dry-box, and yield is 25–26 g. (90%, m.p., 162–163°). (■ *Note. The checkers' product, though analytically and spectroscopically pure, consistently melted at 148–149°.*)

Properties

2,4,6-Trichloro-1,3,5-trimethylborazine is a white, crystalline compound which is sensitive to moisture. The compound is best stored under anhydrous, cold conditions. It is soluble in a number of organic solvents such as benzene and ethyl ether.

The compound can be reduced with sodium borohydride to 1,3,5-trimethylborazine.[3-6] The chlorine can be replaced by fluorine using titanium tetrafluoride or antimony trifluoride to yield 2,4,6-trifluoro-1,3,5-trimethylborazine (m.p., 90.5°).[7] Hexamethylborazine can also be prepared from 2,4,6-trichloro-1,3,5-trimethylborazine.[8]

The infrared spectrum of the compound shows the boron–nitrogen ring frequency at 1392 cm.$^{-1}$. Its ^{1}H n.m.r. spectrum shows a sharp singlet at -3.1 p.p.m. (TMS standard).

References

1. K. Niedenzu and J. W. Dawson, "Boron-Nitrogen Compounds," Springer-Verlag OHG, Berlin, 1965.
2. C. A. Brown and A. W. Laubengayer, *J. Am. Chem. Soc.*, **71**, 3699 (1955).
3. G. H. Dahl and R. Schaeffer, *J. Inorg. Nucl. Chem.*, **12**, 380 (1960).
4. L. F. Hohnstedt and D. T. Haworth, *J. Am. Chem. Soc.*, **82**, 89 (1960).
5. J. Bonham and R. S. Drago, *Inorganic Syntheses*, **9**, 8 (1967).
6. K. Niedenzu and J. W. Dawson, *Inorganic Syntheses*, **10**, 142 (1967).
7. K. Niedenzu, H. Bayer and H. Jenne, *Chem. Ber.*, **96**, 2649 (1963).
8. D. T. Haworth and L. F. Hohnstedt, *J. Am. Chem. Soc.*, **82**, 3860 (1960).

Chapter Two

ORGANOMETALLIC COMPOUNDS

11. CYCLIC-DIOLEFIN COMPLEXES OF PLATINUM AND PALLADIUM

Submitted by D. DREW* and J. R. DOYLE*
Checked by ALAN G. SHAVER†

Recently compounds containing cyclic polyolefins coordinated to platinum or palladium have received considerable attention as a result of the unique bonding found in these compounds and their possible use as intermediates in a variety of reactions. Several methods have been reported for the synthesis of these compounds, and among these procedures the displacement of ethylene[1] from di-μ-chloro-dichlorobis(ethylene)diplatinum(II) and benzonitrile[2] from dichlorobis(benzonitrile)palladium(II) are the most generally applied procedures. Both of these methods involve the preparation of intermediates before the isolation of the product, and in addition these intermediates tend to decompose upon storage.

The procedures described in the subsequent preparations are rapid and the starting materials are the commercially available

* The University of Iowa, Iowa City, Iowa 52240.
† Massachusetts Institute of Technology, Cambridge, Mass. 02139

palladium(II) chloride and hydrated chloroplatinic acid [hydrogen hexachloroplatinate(2−)].

A. PLATINUM COMPOUNDS

The procedure for the preparation of the platinum compounds is an extension of the method described by Kharasch and Ashford.[3] A glacial acetic acid solution of chloroplatinic acid is mixed with the appropriate olefin, and in the ensuing reaction the platinum is reduced from the 4+ oxidation state to the 2+ state. The overall stoichiometry of these reactions is not known; however, the reduction of the platinum is accompanied by the partial oxidation of the olefin.

1. Dichloro(1,5-cyclooctadiene)platinum(II)

$$C_8H_{12} + H_2PtCl_6(H_2O)_x \rightarrow C_8H_{12}PtCl_2*$$

Procedure

In a 125-ml. Erlenmeyer flask 5.0 g. (8.41 mmoles) of hydrated chloroplatinic acid is dissolved in 15 ml. of glacial acetic acid and the solution heated to 75°. Six milliliters of 1,5-cyclooctadiene† is added to the warm solution and the mixture swirled gently, cooled to room temperature, and diluted with 50 ml. of water. The black suspension is stored for one hour at room temperature, and the crude product is collected on a Büchner funnel, washed with 50 ml. of water, and finally 100 ml. of ether. The crude product is suspended in 400 ml. of methylene chloride

---

* ■ *Note.* *Several recently purchased samples of the commercially available hydrated chloroplatinic acid, labeled to contain 40% platinum by weight, were actually analyzed as 32.8% by weight, and the yields were computed on the basis of the latter percentage.* *Caveat emptor.*

† The following hydrocarbons have been successfully substituted for 1,5-cyclooctadiene in this procedure to yield the corresponding dichloro(olefin)platinum(II) derivatives: 1,3,5,7-cyclooctatetraene, dicyclopentadiene (3a,4,7,7a-tetrahydro-4,7-methanoindene), and bicyclo[2.2.1]hepta-2,5-diene (2,5-norbornadiene).

and the mixture heated to the boiling point and kept at this temperature for 5 minutes. The solution is cooled, mixed with 5.0 g. of chromatographic-grade silica gel, and allowed to settle. The supernatant liquid should be colorless; if not, add additional silica gel in 1-g. portions until the solution is clear. The mixture is filtered and the residue washed with two 50-ml. portions of methylene chloride. The methylene chloride solution, approximately 500 ml., is evaporated until the product commences to crystallize, about 75 ml. The hot solution is poured into 200 ml. of petroleum ether (b.p. 60–70°), yielding a finely divided white product. The precipitate is washed with 50 ml. of petroleum ether and dried. Yield is 2.55 g. (80%). *Anal.* Calcd. for $C_8H_{12}PtCl_2$: C, 25.68; H, 3.23. Found: C, 25.73; H, 3.41.

A small amount of product can be recovered by evaporation of the methylene chloride–petroleum ether filtrate. The product can be recrystallized to yield white macroscopic crystals by dissolving the white powder in 150 ml. of boiling methylene chloride and evaporating the solution until crystallization commences.

Properties

Dichloro(1,5-cyclooctadiene)platinum(II) is a white, air-stable solid. The compound is slightly soluble in solvents such as chloroform, acetic acid, sulfolane (tetrahydrothiophene 1,1-dioxide), and nitromethane. It decomposes slowly upon dissolution in dimethyl sulfoxide. The p.m.r. spectrum of the compound in chloroform shows resonances at 4.38τ, $J_{Pt-H} = 65$ Hz., for the olefinic protons and 7.29τ for the methylene protons. The infrared spectrum in Nujol has strong absorption maxima at 1334, 1179, 1009, 871, 834, and 782 cm.$^{-1}$.

2. Dibromo(1,5-cyclooctadiene)platinum(II)

$$C_8H_{12} + H_2PtCl_6(H_2O)_x + 6NaBr \rightarrow C_8H_{12}PtBr_2$$

Procedure

A mixture of 5.0 g. (8.41 mmoles) of hydrated chloroplatinic acid and 6.2 g. (60 mmoles) of sodium bromide suspended in 15 ml. of glacial acetic acid in a 125-ml. Erlenmeyer flask is heated at 75° for 10 minutes. Six milliliters of 1,5-cyclooctadiene* is added to the hot solution yielding a brown solution and a black precipitate containing the impure product. The pure product is isolated in *exactly* the same manner as that described for the preparation of dichloro(1,5-cyclooctadiene)-platinum(II). The clarified methylene chloride solution is very pale yellow. The yield of the very pale yellow product in the form of a finely divided powder was 3.22 g. (83%). *Anal.* Calcd. for $C_8H_{12}PtBr_2$: C, 20.75; H, 2.61. Found: C, 21.21; H, 2.66.

Macroscopic pale yellow crystals can be isolated by the recrystallization procedure described in the preparation of dichloro(1,5-cyclooctadiene)platinum(II).

Properties

Dibromo(1,5-cyclooctadiene)platinum(II) is a very pale yellow, air-stable solid. The solubility of the compound is similar to that of the chloro derivative except that the rate of decomposition in dimethyl sulfoxide is appreciably faster. The p.m.r. spectrum of the compound in chloroform shows resonances at 4.32τ, $J_{Pt-H} = 70$ Hz., for the olefinic protons and 7.43τ for the methylene protons. The infrared spectrum in Nujol has strong absorption maxima at 1334, 1175, 1007, 870, 830, and 780 cm.$^{-1}$.

3. (1,5-Cyclooctadiene)diiodoplatinum(II)

$$C_8H_{12} + H_2PtCl_6(H_2O)_x + 6KI \rightarrow C_8H_{12}PtI_2$$

* The following hydrocarbons have been successfully substituted for 1,5-cyclooctadiene in this procedure to yield the corresponding dibromo(olefin)platinum(II) derivatives: 1,3,4,7-cyclooctatetraene, dicyclopentadiene (3a,4,7,7a-tetrahydro-4,7-methanoindene), and bicyclo[2.2.1]hepta-2,5-diene (2,5-norbornadiene).

Procedure

To a solution of 5.0 g. (8.41 mmoles) of hydrated chloro-platinic acid in 50 ml. of water is added 10 g. (60 mmoles) of potassium iodide followed by 6 ml. of 1,5-cyclooctadiene.*

The brown suspension is stirred magnetically for 30 minutes, and then a solution containing 1.9 g. of *fresh* sodium metabisulfite in 20 ml. of water is added until the solution is colorless. The yellow product is separated on a Büchner funnel and washed with two 50-ml. portions of water and two 100-ml. portions of diethyl ether. The product is redissolved in 300 ml. of boiling methylene chloride, and after cooling, 6.0 g. of chromatographic grade silica gel is added to the solution. The solution is filtered and the silica gel residue washed with two 50-ml. portions of methylene chloride. The combined filtrates and washings are evaporated until crystallization commences (about 75 ml.) and the product recovered by adding the hot solution to 200 ml. of petroleum ether (b.p. 60–70°). The precipitate is washed with 50 ml. of petroleum ether and dried. Yield is 3.60 g. (77%). *Anal.* Calcd. for $C_8H_{12}PtI_2$: C, 17.25; H, 2.17. Found: C, 17.42; H, 2.12.

A small amount of product can be recovered by evaporation of the methylene chloride–petroleum ether filtrate. The product can be recrystallized to yield yellow macroscopic crystals by dissolving the yellow powder in 150 ml. of boiling methylene chloride and evaporating until the solution crystallization commences.

Properties

(1,5-Cyclooctadiene)diiodoplatinum(II) is a yellow, air-stable solid. The compound is slightly soluble in solvents such as

---

* The following hydrocarbons have been successfully substituted for 1,5-cyclo-octadiene in this procedure to yield the corresponding diiodo(olefin)platinum(II) derivatives: 1,3,5,7-cyclooctatetraene and bicyclo[2.2.1]hepta-2,5-diene (2,5-nor-bornadiene).

chloroform, acetic acid, sulfolane (tetrahydrothiophene 1,1-dioxide), and nitromethane. It decomposes upon dissolution in dimethyl sulfoxide. The p.m.r. spectrum of the compound in chloroform shows resonances at 4.24τ, $J_{Pt-H} = 64$ Hz., for the olefinic protons and at 7.63τ for the methylene protons. The infrared spectrum in Nujol has strong absorption maxima at 1334, 1172, 1000, 867, 825, and 776 cm.$^{-1}$.

B. PALLADIUM COMPOUNDS

1. Dichloro(1,5-cyclooctadiene)palladium(II)

$$2HCl + PdCl_2 \rightarrow H_2PdCl_4$$
$$H_2PdCl_4 + C_8H_{12} \rightarrow C_8H_{12}PdCl_2 + 2HCl$$

This compound has been prepared by the reaction of sodium tetrachloropalladate and the olefin[4] or by the displacement of carbon monoxide from $[COPdCl_2]_2$[5] or benzonitrile from $(C_6H_5C{\equiv}N)_2PdCl_2$. Bicyclo[2.2.1]hepta-2,5-diene (2,5-norbornadiene) may be substituted in the following procedure for 1,5-cyclooctadiene to yield (bicyclo[2.2.1]hepta-2,5-diene)dichloropalladium(II).

Procedure

Palladium(II) chloride (2.0 g., 11.3 mmoles) is dissolved in 5 ml. of concentrated hydrochloric acid by warming the mixture. The cool solution is diluted with 150 ml. of absolute ethanol, filtered, and the residue and filter paper washed with 20 ml. of ethanol. To the combined filtrate and washings is added 3.0 ml. of 1,5-cyclooctadiene with stirring. The yellow product precipitates immediately, and after a 10-minute storage is separated and washed with three 30-ml. portions of diethyl ether. Yield is 3.10 g. (96%, based on $PdCl_2$). *Anal.* Calcd. for $C_8H_{12}PdCl_2$: C, 33.66; H, 4.24. Found: C, 34.26; H, 4.39.

The product can be obtained as macroscopic yellow crystals

by dissolving the yellow powder in 200 ml. of boiling methylene chloride and evaporating the hot solution until crystallization commences.

Properties

Dichloro(1,5-cyclooctadiene)palladium(II) is an air-stable yellow solid. The compound is slightly soluble in solvents such as chloroform, sulfolane (tetrahydrothiophene 1,1-dioxide), and nitrobenzene and reacts with dimethyl sulfoxide to yield dichlorobis(dimethyl sulfoxide)palladium(II). The p.m.r. spectrum of the compound in chloroform shows resonances at 3.68τ for the olefinic protons and at 7.31τ for the methylene protons. The infrared spectrum in Nujol has strong absorption maxima at 1489, 1419, 1337, 1088, 999, 867, 825, 794, 768, 325, and 295 cm.$^{-1}$.

2. Dibromo(1,5-cyclooctadiene)palladium(II)

$$2HCl + PdCl_2 \rightarrow H_2PdCl_4$$
$$H_2PdCl_4 + 4NaBr \rightarrow 4NaCl + H_2PdBr_4$$
$$H_2PdBr_4 + C_8H_{12} \rightarrow C_8H_{12}PdBr_2 + 2HBr$$

The previously reported[4] procedure, which is a technique of general utility, involves the reaction of dichloro(1,5-cycloocta-diene)palladium(II) with a solution of sodium bromide in acetone. The following procedure gives comparable yields and eliminates the need for the preparation of the dichloro intermediate. Bicyclo[2.2.1]hepta-2,5-diene (2,5-norbornadiene) may be substituted in the following procedure for the 1,5-cyclooctadiene to yield (bicyclo[2.2.1]hepta-2,5-diene)dibromopalladium(II).

Procedure

Two grams of palladium(II) chloride (11.3 mmoles) is dissolved in 5 ml. of concentrated hydrochloric acid by warming

the mixture.   A solution containing 4.65 g. (45.2 mmoles) of
sodium bromide in 7.0 ml. of water is added to the palladium
chloride solution and the mixture warmed to 50° for about
5 minutes.   The mixture is diluted with 50 ml of absolute
ethanol and after cooling for 15 minutes is filtered.   The flask
and filter paper are washed with three 10-ml. portions of 75%
aqueous ethanol to remove the residual palladium salts.   To
the combined filtrate and washings is added 3.0 ml. of 1,5-cyclo-
octadiene, and the solution is mixed by swirling the flask.   The
orange product precipitates immediately and after settling for
10 minutes is collected on a Büchner funnel.   The product
remaining in the flask is transferred to the funnel with the aid of
50 ml. of water, and the collected product is washed with an
additional 50 ml. of water, followed by two 100-ml. portions of
ether.   Yield is 3.96 g. (93%, based on $PdCl_2$).   *Anal.* Calcd.
for $C_8H_{12}PdBr_2$: C, 25.66; H, 3.23.   Found: C, 26.85; H, 3.17.

The product may be obtained as macroscopic orange crystals
by dissolving the orange powder in 150 ml. of boiling methylene
chloride and evaporating the hot solution until crystallization
commences.

Properties

Dibromo(1,5-cyclooctadiene)palladium(II) is an air-stable,
orange solid.   The compound is slightly soluble in solvents such
as chloroform, nitrobenzene, and sulfolane (tetrahydrothiophene 1,1-dioxide), and reacts rapidly with dimethyl sulfoxide
to yield dibromobis(dimethyl sulfoxide)palladium(II).   The
p.m.r. spectrum of the compound in chloroform shows resonances
at 3.58τ for the olefinic protons and 7.40τ for the methylene
protons.   The infrared spectrum in Nujol has strong absorption
maxima at 1472, 1417, 1333, 1172, 1083, 992, 905, 864, 823, 787,
764, 678, 310, 265, 221, 213, 178, and 126 cm.$^{-1}$.

References

1. J. Anderson, *J. Chem. Soc.*, **1936**, 1042.
2. M. S. Kharasch, R. C. Seyler, and F. R. Mayo, *J. Am. Chem. Soc.*, **60**, 822 (1938).
3. M. S. Kharasch and T. Ashford, *ibid.*, **58**, 1733 (1936).
4. J. Chatt, L. M. Vallarino, and L. M. Venanzi, *J. Chem. Soc.*, **1957**, 3413.
5. E. O. Fischer and H. Werner, *Chem. Ber.*, **93**, 2075 (1960).

12. CATIONIC DIENE COMPLEXES OF PALLADIUM(II) AND PLATINUM(II)

Submitted by D. A. WHITE*
Checked by J. R. DOYLE† and H. LEWIS†

The palladium(II) and platinum(II) ions form stable complexes with a variety of chelating diolefins.[1-5] These may be either neutral or cationic in character. The preparative routes to the former type are, in general, well-documented. The species (I) through (IV) include all the presently known cationic species, and the preparation of each type is discussed and exemplified.

(I) $[(\text{diene})M(\text{ch})]^{+}$

    diene = 1,5-cyclooctadiene, 2,5-norbornadiene, or dicyclopentadiene (3a,4,7,7a-tetrahydro-4,7-methanoindene)

    ch = conjugate base of acetylacetone (2,4-pentanedione), benzoylacetone (1-phenyl-1,3-butanedione) or dibenzoylmethane (1,3-diphenyl-1,3-propanedione)

(II) $[(\text{diene})M(h^{5}\text{-}C_{5}H_{5})]^{+}$

    diene = 1,5-cyclooctadiene or 1,2,3,4-tetraphenyl-1,3-cyclobutadiene

(III) $[(\text{cyclooctadiene})Pd(h^{5}\text{-}C_{3}H_{4}R)]^{+}$

    R = H or 1-Me

* The Monsanto Company, 800 N. Lindbergh Blvd., St. Louis, Mo. 63166.
† The University of Iowa, Iowa City, Iowa 52240.

(IV) [(cyclooctadiene)Pt(L)X)]$^+$
L = *tert*-phosphine or arsine
X = halogen

A. COMPLEXES OF TYPE (I)

These have been obtained by electrophilic attack on ene-yl complexes (equation a):

a. (Diene—Y)M(ch) + Ph$_3$C$^+$ → [(diene)M(ch)]$^+$ + Ph$_3$CY
              or H$^+$ or HY
           Y = CH(COMe)$_2$, OMe, OH

This reaction demonstrates well the expected susceptibility of these ene-yl complexes to electrophilic attack, but a more convenient preparative method is shown in equation (b).

b. (Diene)MCl$_2$ + 2AgBF$_4$ + Hch →
                    [(diene)M(ch)]$^+$ + 2AgCl + HBF$_4$

These reactions are of short duration and go in high yield. They are exemplified by the preparations of the cations (I, diene = 1,5-cyclooctadiene, M = Pd or Pt, ch = conjugate base of acetylacetone).

Procedure

**1. (1,5-Cyclooctadiene)(2,4-pentanedionato) palladium(II)
Tetrafluoroborate**

$$K_2[PdCl_4] + diene → [Pd(diene)Cl_2] + 2KCl$$
$$[Pd(diene)Cl_2] + Hch + 2AgBF_4 →$$
$$[Pd(diene)ch]BF_4 + 2AgCl + HBF_4$$

A mixture of potassium tetrachloropalladate(II) (3.26 g., 10.0 mmoles), water (125 ml.), and 1,5-cyclooctadiene (2.0 ml.) is shaken vigorously for 15 minutes, during which time a yellow solid forms. This is filtered, washed with water (three times with 30 ml.), and dried *in vacuo* overnight to give dichloro-(1,5-cyclooctadiene)palladium(II) (2.69 g.).

The diene complex is transferred to a 100-ml. Erlenmeyer flask and methylene chloride (50 ml.) added. Silver tetrafluoroborate (3.90 g., 20.0 mmoles.)* is added and the mixture stirred for 15 minutes. At this stage the methylene chloride is pale yellow and there is a heavy yellow precipitate. Acetylacetone (1.5 ml.) is added and the mixture stirred for 1 minute, after which time the solution becomes deep orange and the precipitate off-white. The mixture is filtered and the solid residue washed with methylene chloride (two times with 15 ml.). Ether (300 ml.) is added to the combined filtrate and washings. The yellow precipitate which forms is filtered and washed with ether (three times with 30 ml.). After sucking dry it is dissolved in methylene chloride (50 ml.). This brownish, somewhat opaque solution is filtered through a tightly packed plug of cotton in a powder funnel.† Ether (150 ml.) is added in small portions to the clear orange filtrate, and the yellow precipitate which forms is filtered and washed with ether (three times with 30 ml.). Drying *in vacuo* affords the product as shiny deep yellow crystals. Yield is 3.08 g. (77% based on K_2PdCl_4 used‡). *Anal.* Calcd. for $C_{13}H_{19}BF_4O_2Pd$: C, 39.0; H, 4.8; F, 19.0; Pd, 26.6. Found: C, 39.0; H, 4.8; F, 18.8; Pd, 26.4. The product is stored in a dark bottle at 5°.

2. (1,5-Cyclooctadiene)(2,4-pentanedionato)platinum(II) Tetrafluoroborate

$$[PtBr_2(diene)] + HC_5H_7O_2 + 2AgBF_4 \rightarrow$$
$$[Pt(C_5H_7O_2)(diene)]BF_4 + 2AgBr + HBF_4$$

---

* Silver tetrafluoroborate is very hygroscopic and should be weighed in a nitrogen-flushed dry-box. However, the reaction may be carried out without precautions against ingress of moisture. Care should also be taken to avoid staining of hands and clothing by solutions of silver tetrafluoroborate which are formed when it comes into contact with moist air.

† The stem of the powder funnel had an internal diameter of 9 mm. and the plug of cotton wool used was *ca.* 40 mm. in length. This was sufficient to remove traces of finely divided metal. If the filtrate is not clear, it must repeatedly be filtered until it is so. It is also advisable to wash the cotton filter with two 10-ml. portions of CH_2Cl_2 to remove all product.

‡ Checkers obtained only 1.65 g. (44%).

Dibromo(1,5-cyclooctadiene)platinum is prepared* and 4.4 g. transferred to a 250-ml. Erlenmeyer flask and methylene chloride (100 ml.) and silver tetrafluoroborate (3.9 g., 20.0 mmoles) added. This mixture is stirred for 30 minutes, after which time acetylacetone (1.5 ml.) is added and stirring continued for a further 5 minutes. The mixture is then filtered, the off-white residue being washed with methylene chloride (two times with 10 ml.). Ether (400 ml.) is added to the combined filtrate and washings giving a white precipitate which is filtered and washed with ether (four times with 50 ml.). The white solid is sucked dry and then dissolved in methylene chloride (50–100 ml.). The solution is filtered through a tightly packed plug of cotton wool in a powder funnel and ether (250 ml.) added in portions to the filtrate. The white precipitate is filtered and washed with ether (four times with 50 ml.). Drying *in vacuo* afforded the product as shiny white platelets. Yield is 4.4 g. (77% based on $C_8H_{12}PtBr_2$). *Anal.* Calcd. for $C_{13}H_{19}BF_4O_2Pt$: C, 31.9; H, 3.9; F, 15.5; O, 6.5; Pt, 39.9. Found: C, 31.9; H, 3.9; Pt, 40.3.

Properties

The complexes (1,5-cyclooctadiene)(2,4-pentanedionato)-palladium(II) and platinum(II) tetrafluoroborate are air-stable solids, soluble in polar organic solvents such as chloroform, methylene chloride, acetonitrile, acetone, or methanol but insoluble in nonpolar solvents such as alkanes, benzene, or ether. Their solutions in acetone have conductivities typical of 1:1 electrolytes. Their proton magnetic resonance spectra (in $CDCl_3$ solutions, internal tetramethylsilane reference at 60 MHz.) show peaks due to coordinated cyclooctadiene at 3.78 and 6.7–7.4τ (Pd) and at 4.25 and 6.9–7.6τ (Pt) and due to the chelated β-diketone at 4.39 and 7.88τ (Pd) and at 4.15 and 7.81τ (Pt) with the expected area ratios. In the spectrum of the platinum compound coupling with the [195]Pt isotope (33%

---

* Dichloro(cyclooctadiene)platinum may also be used as an intermediate.

abundance) is observed for the bands at 4.25 [J_{Pt-H}, 60 Hz.], 4.39 [J_{Pt-H}, 9 Hz.], and 7.81τ [J_{Pt-H} 4.5 Hz.].

Neither compound shows a definite melting point, the palladium compound decomposing above 100° and the platinum above 185°. The platinum compound can be stored in air at room temperature, but the palladium decomposes slowly under these conditions and is best stored in a refrigerator in a brown bottle. In this way it may be kept for at least several months without decomposition.

The chemistry of these compounds has been only briefly examined. They react with methanol in the presence of a base giving the ene-yl complexes $(C_8H_{12}$—OMe$)MC_5H_7O_2$.[1] With triphenylphosphine the olefin may be displaced, and in this way $[(Ph_3P)_2PtC_5H_7O_2]BF_4$ is obtained conveniently.[1]

B. COMPLEXES OF TYPE (II)

These have previously been obtained by electrophilic attack on ene-yl complexes [equation (a); Y = $CH(CO_2Me)_2$, OMe; ch = η-C_5H_5; diene = 1,5-cyclooctadiene][1] or by reaction of the compounds (diene)MBr_2 with η-$C_5H_5Fe(CO)_2Br$ (diene = 1,5-cyclooctadiene or 1,2,3,4-tetraphenyl-1,3-cyclobutadiene).[2] An example of the former method is given in which the methoxy-cyclooctenyl derivative is used as the substrate and tetrafluoroboric acid as the electrophile. The substrate is conveniently prepared and used without isolation, and in this way the reaction takes only a few hours, starting with dichloro(1,5-cyclooctadiene)palladium, prepared as described above.

Procedure

1. (1,5-Cyclooctadiene)cyclopentadienyl palladium(II)
   Tetrafluoroborate

$2[Pd(diene)Cl_2] + 2CH_3OH + Na_2CO_3 \rightarrow$
$\qquad\qquad [\{(CH_3O\text{-eneyl})Pd\}_2Cl_2] + 2NaCl + H_2O + CO_2$
$[\{(CH_3O\text{-eneyl})Pd\}_2Cl_2] + 2TlC_5H_5 + 2HBF_4 \rightarrow$
$\qquad\qquad 2[Pd(diene)(h^5\text{-}C_5H_5)]BF_4 + 2TlCl + 2CH_3OH$

The diene complex, dichloro(1,5-cyclooctadiene)palladium(II) (2.85 g., 10.0 mmoles), and sodium carbonate (0.53 g., 5.0 mmoles) are suspended in methanol (50 ml.), and the suspension is stirred for 45 minutes. It is then filtered and the off-white residue washed with water (three times with 50 ml.) followed by methanol (two times with 10 ml.). Drying *in vacuo* gives di-μ-dichlorobis(8-methoxy-4-cycloocten-1-yl)dipalladium[3] (2.6 g.). This is suspended in acetone (75 ml.), and cyclopentadienylthallium[4] (2.65 g., 10.0 mmoles) is added. The mixture is stirred for 10 minutes, and then the precipitated thallium chloride is filtered. The thallium chloride is washed with acetone (two times with 10 ml.). To the combined orange-colored filtrate and washings, which contain (8-methoxy-4-cycloocten-1-yl)cyclopentadienylpalladium(II), is added 48% aqueous tetrafluoroboric acid (3.0 ml.). The solution turns deep violet instantly. Addition of ether (250 ml.) gives a violet precipitate which is separated by filtration or decantation and washed with ether (three times with 50 ml.). After drying in air the precipitate is dissolved in methylene chloride (50–100 ml.), and the solution is filtered through a tightly packed plug of cotton in a powder funnel and the plug washed with 50 ml. of CH_2Cl_2. Addition of ether (200 ml.) to the filtrate precipitates the product, which, after filtration, washing with ether, and drying *in vacuo*, is obtained as a violet powder. Yield is 3.41 g. (93% based* on $C_8H_{12}PdCl_2$). *Anal.* Calcd. for $C_{13}H_{17}BF_4Pd$: C, 42.6; H, 4.7; Pd, 29.0. Found: C, 42.5; H, 4.18.

Properties

(Cycloocta-1,5-diene)cyclopentadienylpalladium(II) tetrafluoroborate is an air-stable solid, soluble in polar solvents such as chloroform, methylene dichloride, acetone, or methanol and insoluble in alkanes, benzene, or ether. Its solution in acetone displays a conductivity characteristic of a 1:1 electrolyte.[1] Its

* Checkers obtained only a 25% yield.

proton magnetic resonance spectrum (in $CDCl_3$ solution, at 60 MHz., with internal tetramethylsilane reference) shows bands at 3.93 (cyclopentadienyl protons) and at 3.75 and 7.2–7.5τ (olefin protons) in the expected area ratios. Its infrared spectrum shows a weak band at 1506 cm.$^{-1}$ attributed to the coordinated-double-bond stretching mode; other bands characteristic of the cyclopentadienyl ligand and of the tetrafluoroborate anion are also observed.

C. COMPLEXES OF TYPE (III)

These compounds have been obtained by the addition of cyclooctadiene to an equimolar mixture of h^3-allyl or h^3-crotyl-(2,4-pentanedionato)palladium(II) and tetrafluoroboric acid in methylene chloride–ether solution.[1] If silver tetrafluoroborate is on hand, the slight modification described below obviates the need to prepare the β-diketonate complex as an intermediate.

Procedure

1. h^3-Allyl(1,5-cyclooctadiene)palladium(II) Tetrafluoroborate

$$[\{(C_3H_5)Pd\}_2Cl_2] + 2(\text{diene}) + 2AgBF_4 \rightarrow$$
$$2[Pd(C_3H_5)(\text{diene})]BF_4 + 2AgCl$$

The dimer di-μ-dichloro-bis(h^3-allyl)dipalladium* (1.83 g., 5.0 mmoles) and silver tetrafluoroborate (1.95 g., 10.0 mmoles) in methylene chloride (50 ml.) are stirred for 15 minutes. 1,5-Cyclooctadiene (2.0 ml.) is added and stirring continued for a further 2 minutes. The mixture is filtered and the residue washed with methylene chloride (two times with 10 ml.). Ether (150 ml.) is added to the combined filtrate and washings to give a white or greyish precipitate. This is filtered, washed with ether (three times with 50 ml.), and dried in air. The solid is dissolved in methylene chloride (75 ml.), and this solution is

---

* Prepared in 66% yield by the procedure of W. T. Dent, R. Long, and A. J. Wilkinson, *J. Chem. Soc.*, **1964**, 1585.

repeatedly filtered through a tightly packed plug of cotton until the solution is absolutely clear. The plug is washed with 20 ml. of CH_2Cl_2. Addition of ether (120 ml.) precipitates the product, which after filtration, washing with ether, and drying *in vacuo* is obtained as an off-white solid. Yield is 2.63 g. (77%). *Anal.* Calcd. for $C_{11}H_{17}BF_4Pd$: C, 38.6; H, 5.0; F, 22.2; Pd, 31.0. Found: C, 38.3; H, 5.0.

Properties

h^3-Allyl(1,5-cyclooctadiene)palladium(II) tetrafluoroborate is a white solid; however, due to the difficulty of obtaining it absolutely free from traces of palladium metal, it is usually greyish in appearance. These small traces of metal do not accelerate its decomposition, as sometimes happens with other palladium complexes, and it is stable for several months when stored in this state. Its solubility properties are similar to those of the compounds described above. It is a 1:1 electrolyte in acetone, and its proton magnetic resonance spectrum shows bands at *ca.* 3.9 (complex), 5.02 (doublet, $J = 7$ Hz.), and 6.17 (doublet, $J = 13$ Hz.) for the h^3-allylic protons and at 3.68 and 7.38τ for the coordinated olefin in the expected area ratios. A band near 1560 cm.$^{-1}$ in its infrared spectrum is attributed to the stretching vibration of the coordinated double bonds.

D. COMPLEXES OF TYPE (IV)

These complexes have been obtained by the somewhat tedious route shown below:

$$C_8H_{12}PtX_2 \xrightarrow{TlC_5H_7O_2} (C_8H_{12}{-}C_5H_7O_2)PtC_5H_7O_2$$
$$(V)$$

$$\downarrow 1 \text{ mole HX}$$

$$(C_8H_{12}{-}C_5H_7O_2)Pt(L)X \xleftarrow{L} [(C_8H_{12}{-}C_5H_7O_2)PtX]_2$$
$$(VII) \qquad\qquad\qquad (VI)$$

$$\downarrow Ph_3C^+$$

$$[C_8H_{12}Pt(L)X]^+ \qquad\qquad X = \text{halogen}$$

However, this is the only route available to these compounds at the present time. The intermediate (VII) must be isolated and purified. In the example given below L = Ph₃As, X = Cl.

Procedure

1. [8-(1-Acetylacetonyl)-4-cycloocten-1-yl]chloro(triphenylarsine)-platinum(II) (VII)[5]

[PtCl₂(diene)] + 2TlCHAc₂ → [Pt(Ac₂CH-eneyl)(Ac₂CH)] + 2TlCl
2[Pt(Ac₂CH-eneyl)(Ac₂CH)] + 2HCl →
[{Pt(Ac₂CH-eneyl)Cl}₂] + 2H₂CAc₂
[{Pt(Ac₂CH-eneyl)Cl}₂] + 2Ph₃As → 2[Pt(Ac₂CH-eneyl)Cl(Ph₃As)]

Dichloro(1,5-cyclooctadiene)platinum(II)   (2.32   g.,   50 mmoles), prepared as described above, and (2,4-pentanedio-nato)thallium (3.03 g., 10.0 mmoles) are stirred in chloroform (200 ml.) for one hour. The thallium chloride formed is removed by filtration. Evaporation gives a yellowish oily solid, which is taken up in ether (100 ml.). Traces of thallium salts are removed from this slightly opaque solution by filtration through a short column (*ca.* 10 cm.) of silica gel. The solute is washed through the column with more ether (250 ml.). Evaporation of the ether solution gives the intermediate (V) as a yellow or colorless oil, which crystallizes on standing. The oil is dissolved in acetone (100 ml.) and hydrochloric acid (5.0 ml. of 1*N*) is added to the magnetically stirred solution. A precipitate begins to form. The mixture is evaporated to dryness to give the dimer (VI) as a greyish-yellow solid. This is suspended in methylene chloride (100 ml.). When triphenyl-arsine (1.53 g., 5.0 mmoles) is added, all or most of the sus-pended material dissolves after shaking. The solution is filtered and then evaporated to dryness. The white residue is washed with ether (four times with 50 ml.) and recrystallized by dissolving in chloroform and precipitating with ether. In

this way, the product (VII) is obtained.   Yield is 2.64 g. (71%).*
Anal. Calcd. for $C_{31}H_{34}AsClO_2Pt$: C, 50.0; H, 4.61.   Found:
C, 49.5; H, 4.4.

2. Chloro(1,5-cyclooctadiene)(triphenylarsine)platinum(II) Tetrafluoroborate

$$[PtCl(Ac_2CH\text{-eneyl})(Ph_3As)] + Ph_3CBF_4 \rightarrow$$
$$[PtCl(diene)(Ph_3As)]BF_4 + Ph_3CCHAc_2$$

The intermediate (VII) (1.48 g., 2.0 mmoles) is dissolved in
methylene chloride (50 ml.), and trityl tetrafluoroborate
(660 mg., 2.0 mmoles) is added.   The clear yellow solution is
allowed to stand for 5 minutes; then ether (200 ml.) is added.
This precipitates an oily white solid.   The mixture is allowed
to stand until the supernatant liquid is clear (*ca.* 1.5 hours).
The supernatant is then poured off and the solid clinging to the
walls of the flask is washed with ether (three times with 50 ml.).
It is then dried in a stream of nitrogen and recrystallized twice
by dissolving in methylene chloride (*ca.* 50 ml.) and precipitating
with ether (*ca.* 200 ml.).   This gives the product as a white
solid.   Yield is 1.07 g. (73%).†   *Anal.* Calcd. for $C_{26}H_{27}$-
$AsBClF_4Pt$: C, 42.7; H, 3.7; As, 10.2.   Found: C, 42.5; H, 3.6;
As, 10.0.

Properties

Chloro(1,5-cyclooctadiene)(triphenylarsine)platinum(II) tetrafluoroborate is a white, air-stable solid.   It is sparingly soluble
in polar organic solvents and insoluble in nonpolar ones.   A
10^{-3} *M* solution in acetone shows a conductivity characteristic
of a 1:1 electrolyte.[1]   Its proton magnetic resonance spectrum
(in liquid sulfur dioxide solution, at 60 MHz., with internal
tetramethylsilane reference) shows bands due to the coordinated olefin at 3.59 [two protons; J_{Pt-H}, 55 Hz.], 5.01 [two

* Checkers obtained 1.94 g. (52%).
† Checkers obtained 0.36 g. (25%).

protons; J_{Pt-H} = 65 Hz.], and 6.8–7.8 (eight protons) and due to the arsine at about 2.4τ (15 protons).

References

1. B. F. G. Johnson, T. Keating, J. Lewis, M. S. Subramanian, and D. A. White, *J. Chem. Soc.* (A), **1969**, 1793.
2. P. M. Maitlis, A. Efraty, and M. L. Games, *J. Am. Chem. Soc.*, **87**, 719 (1965); *J. Organometallic Chem.*, **2**, 284 (1964).
3. R. G. Schultz, *ibid.*, **6**, 435 (1966).
4. H. Meister, *Angew. Chem.*, **69**, 533 (1957); J. M. Birmingham, *Advan. Organometallic Chem.*, **2**, 373 (1964).
5. B. F. G. Johnson, J. Lewis, and M. S. Subramanian, *J. Chem. Soc.* (A), **1968**, 1993.

13. TRICHLORO-, TRIMETHYL-, AND TRIFLUOROSILYLCOBALT TETRACARBONYL
[*Tetracarbonyl(trichloro-, trifluoro-, and Trimethyl-silyl)cobalt Complexes*]

Submitted by DAVID L. MORRISON* and ARNULF P. HAGEN*
Checked by N. VISWANATHAN† and C. H. VAN DYKE‡

The syntheses of several compounds having the formulas $R_3SiCo(CO)_4$ (R = Cl, C_2H_5O, C_2H_5)[1] and of $H_3SiCo(CO)_4$[2] were reported in 1965. The procedures given here for three specific compounds are representative of a synthetic method which has been used by other workers to prepare compounds such as $Cl_3SiMn(CO)_5$,[3] $(C_6H_5)_3SiMn(CO)_5$,[4] $Cl_3SiMo(CO)_3$-C_5H_5,[3] $(Cl_3Si)_2Fe(CO_3)$,[3] and $Cl_3SiNi(CO)C_5H_5$.[3] The general scheme for reaction consists of the interaction of a silicon hydride with a transition-metal carbonyl compound.

In the case of $Co_2(CO)_8$ interaction, it has been clearly established[5,6] that complete reaction takes place by a two-step inter-

* University of Oklahoma, Norman, Okla. 73069.
† Pennsylvania State University, Uniontown Campus, Uniontown, Pa. 15401.
‡ Carnegie-Mellon University, Pittsburgh, Pa. 15213.

action. The first step results in the formation of $HCo(CO)_4$ and $R_3SiCo(CO)_4$:

$$R_3SiH + Co_2(CO)_8 \rightarrow R_3SiCo(CO)_4 + HCo(CO)_4$$

The second step results in the generation of hydrogen gas and the formation of another molecule of $R_3SiCo(CO)_4$:

$$R_3SiH + HCo(CO)_4 \rightarrow R_3SiCo(CO)_4 + H_2$$

The cobalt–silicon compounds described here undergo reactions with protonic molecules such as hydrogen chloride, Lewis bases such as trimethylamine, and oxidizing agents.

Each synthesis is carried out in a glass pressure vessel similar to that illustrated in Fig. 2. This type of reactor was made from a Fisher and Porter (Warminster, Pa.), 4-mm., quick-opening angle valve (catalog no. 795-005-004) and a length of heavy-walled borosilicate glass tubing (18 mm. i.d., 26 mm. o.d.) cut to give the desired volume.* This vessel has been found to be

* The checkers used a 35-ml. vessel.

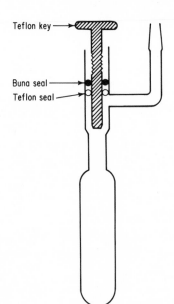

Fig. 2. *Pressure vessel for the synthesis of substituted silylcobalt carbonyls.*

suitable for reaction mixtures having an autogenous pressure of up to 50 atmospheres. The vessel when under pressure is normally manipulated behind a safety shield with the operator wearing a leather glove. Volatile reactants are purified by the use of normal vacuum-line procedures and then condensed directly into the reactor.* At the end of the reaction period, if the reactor is cooled to liquid-nitrogen temperature and then opened to a manometer, noncondensable material (presumed to be H_2 and CO of decomposition) will be observed.

■ *Caution. The silyl compounds described in the following preparations react with oxygen and water vapor. All non-vacuum-line manipulations of these materials require the use of inert atmospheres (deoxidized nitrogen† or carbon monoxide) and an efficient fume hood. The toxicity of these compounds is unknown, but they are assumed to be extremely dangerous as are most transition-metal carbonyls.*

A. TETRACARBONYL(TRICHLOROSILYL)COBALT

$$2Cl_3SiH + Co_2(CO)_8 \rightarrow 2Cl_3SiCo(CO)_4 + H_2$$

Procedure

Commercial trichlorosilane,‡ $HSiCl_3$ (about 8 g.), is placed in a glass tube which can be attached to the vacuum system through a stopcock. The $HSiCl_3$ is then cooled to liquid-nitrogen temperature ($-196°$), and then the tube is evacuated. The stopcock is then closed and the nitrogen bath removed. After the sample has melted, it is solidified at $-196°$ and the tube evacuated again. The freezing and melting process is continued until

* A suitable vacuum line is depicted by W. L. Jolly and J. E. Drake, *Inorganic Syntheses*, **7**, 35 (1963).

† Liquid nitrogen is required for the vacuum-line manipulations described in this report. Gaseous nitrogen vented from a standard· liquid-nitrogen cylinder is a convenient source of deoxidized nitrogen.

‡ Available as Silane A-19 from Union Carbide Corporation.

the sample has been freed of noncondensable impurities (air). The contents of the tube are then warmed slowly to room temperature while distilling through a $-132°$ trap (n-pentane slush) into a $-196°$ trap. The material which stops in the $-132°$ trap (HSiCl₃) is cooled to $-196°$ and then again warmed slowly to room temperature and allowed to distill through a $-132°$ trap into a $-196°$ trap. This process is repeated until no HCl is observed in the $-196°$ trap. The process may be represented by:

$$RT \sim -132° \sim -196°$$
$$\downarrow \text{(HCl)}$$
$$RT \sim -132° \sim -196°$$
$$\downarrow \text{(HCl)}$$

repeated until no material
passes into the $-196°$ trap

where RT means to warm the material slowly from $-196°$ to room temperature. The process is carried out without pumping unless noted. The material in the $-196°$ trap (HCl) is discarded. The trichlorosilane in the $-132°$ trap will have an infrared spectrum[7] and a molecular weight consistent with the literature.

A 2.0-g. sample of powdered dicobalt octacarbonyl is placed in a 25-ml. glass reactor in a nitrogen-filled glove bag.* The reactor is evacuated; then 5 g. of trichlorosilane is condensed from the vacuum system into the reactor which has been cooled to $-196°$. The Teflon stopcock is closed, the reaction vessel allowed to warm to room temperature, and then the reactants are permitted to stand for 24 hours. Cool the reactor to $-42°$ (diethyl ketone slush), open the Teflon stopcock, and remove the excess silane and noncondensable substances into the vacuum system with pumping. The remaining dry, solid material is then transferred in a nitrogen (or carbon monoxide) filled glove bag to a sublimation apparatus. The solid is then sublimed *in vacuo*

---

* I²R Company, Cheltenham, Pa. 19012.

twice at 30° using a cold finger at −5° to give tetracarbonyl-(trichlorosilyl)cobalt.* The product is then scraped from the cold finger in an inert atmosphere and placed in bottles with tight-fitting caps. The compound which is recovered in a 64% yield based on the cobalt carbonyl, $Co_2(CO)_8$, employed should be stored in a freezer (−10°).

Properties

The tetracarbonyl(trichlorosilyl)cobalt obtained will be a light yellow solid with an observed melting point of 44.5–45.0° (literature,[5] 44°). The infrared spectrum in the 3000–600-cm.$^{-1}$ region exhibits only three major absorptions. In CCl_4 solution[8] peaks due to carbonyl stretching vibrations are observed at 2118(m), 2066(s), and 2039(vs) cm.$^{-1}$, and in the gas phase[9] these bands are found at 2128(m), 2073(s), and 2049(vs) cm.$^{-1}$. The presence of impurities can be detected by their infrared absorptions[10] outside of the CO stretching region. Gaseous $Co_2(CO)_8$ has characteristic bonds at 1876(s) and 641(s) cm.$^{-1}$, and gaseous $HCo(CO)_4$ absorbs at 1934(m) and 704(vs) cm.$^{-1}$.

B. TETRACARBONYL(TRIMETHYLSILYL)COBALT

$$2(H_3C)_3SiH + Co_2(CO)_8 \rightarrow 2(H_3C)_3SiCo(CO)_4 + H_2$$

Two grams of powdered dicobalt octacarbonyl are placed in a 20-ml. glass pressure reactor in a nitrogen atmosphere. The reactor is then evacuated and cooled to −196°. Then 4 g. of trimethylsilane, $(H_3C)_3SiH$,† are condensed into the reactor, and the Teflon stopcock is closed. The reactor is then allowed to warm to room temperature. After 24 hours the reactor is cooled to −196°, the stopcock opened, and any H_2 and CO are

---

* ■ *Note.* *The use of "cool" tapwater in the sublimation apparatus is not recommended. A suitable circulatory or other system which uses ice water or, better yet, an ice–salt-water mixture should be used.*

† Available from Pierce Chemical Company, P. O. Box 117, Rockford, Ill. 61105, or easily prepared from $LiAlH_4$ and $(H_3C)_3SiCl$.

removed with pumping. Then it is warmed to 0°, and the volatile materials in the reactor are allowed to distill from the reactor into a trap at −196°. The volatile material which collects in the −196° trap is warmed to 0° and allowed to distill with pumping through cold traps at −42° [diethyl ketone (3-pentanone) slush] and −196°. The clear, rose-colored crystals in the −42° trap are the desired product. Yield is 38% based on $Co_2(CO)_8$ employed. Single crystals melt at 51.4° with some decomposition; however, larger bulk samples melt somewhat lower due to contamination from decomposition products.[6]

Properties

Tetracarbonyl(trimethylsilyl)cobalt is a pink solid with a melting point of 51–53°.[6] The infrared spectrum of the gaseous compound contains characteristic carbonyl bands at 2100(m), 2041(s), and 2009(vs) cm.$^{-1}$, as well as a strong methyl symmetric deformation band at 1252 cm.$^{-1}$ in addition to other absorptions.[6]

C. TETRACARBONYL(TRIFLUOROSILYL)COBALT

$$HSiCl_3 \xrightarrow{SbF_3} HSiF_3 \xrightarrow{Co_2(CO)_8} H_2 + F_3SiCo(CO)_4$$

Procedure

In a nitrogen-filled glove bag 10 g. of antimony trifluoride* are placed into a 30-ml. glass reactor which is then evacuated and pumped on for 24 hours to provide a final drying at room temperature. The reactor is then cooled to liquid-nitrogen temperature, and 3 g. of hydrogen-chloride-free trichlorosilane is added to the reactor. The Teflon stopcock is then closed

---

* Antimony trifluoride suitable for use in this synthesis is available from the Ozark-Mahoning Company, 1870 South Boulder, Tulsa, Okla. 74119.

and the mixture warmed to $-42°$ (diethyl ketone slush) where it is held for 2 hours. The reactor is then cooled to liquid-nitrogen temperature and opened to a manometer. If any noncondensable gas is observed (presumably H_2), it is removed with pumping. The vessel is then warmed to room temperature, and the volatile components (unreacted $HSiCl_3$, $HSiF_3$, and perhaps a trace of SiF_4) are condensed, with occasional pumping, in two $-196°$ traps in series. The substances in the $-196°$ traps are combined and then distilled at room temperature through a trap at $-132°$ (*n*-pentane slush) into a trap at $-196°$. Trichlorosilane collects in the $-132°$ trap; $HSiF_3$ and SiF_4 collect in the $-196°$ trap. The distillation is repeated on the fraction which collects at $-196°$ until all of the trichlorosilane has been removed. The distillation scheme may be shown as:

$$RT \sim -132° \sim -196°$$
$$\quad (\text{HSiCl}_3) \qquad \downarrow$$
$$\qquad RT \sim -132° \sim -196°$$
$$\qquad\qquad \downarrow$$
$$\qquad\qquad \text{repeated until all of the HSiCl}_3$$
$$\qquad\qquad \text{has been removed from the HSiF}_3$$

Quite often additional trifluorosilane can be obtained by simply returning the unreacted silane to the reaction vessel and allowing additional reaction time. A single reaction period yields about 1.5 g. (65% based on $HSiCl_3$ employed) of product having an observed molecular weight of 86.9 (calcd. 86.1) and an infrared spectrum consistent with pure trifluorosilane.[11] However, on some occasions the product is contaminated with silicon tetrafluoride due to a reaction of the antimony compounds with the trifluorosilane. The silicon tetrafluoride impurity does not interfere with the preparation of tetracarbonyl(trifluorosilyl)cobalt.

Next 2 g. of dicobalt octacarbonyl is placed in a 20-ml. glass reactor in a nitrogen-filled glove bag. After evacuation of the reactor, 1.8 g. of trifluorosilane is condensed and the Teflon stopcock closed. The reaction is allowed to proceed for 24 hours at room temperature. A dark liquid should be observed at the bottom of the reactor after the reaction.

Cool the reactor to $-196°$ and remove the noncondensable materials by pumping. Allow the reactor to warm to $0°$ and remove the excess trifluorosilane, tetracarbonyl(trifluorosilyl)-cobalt, and any cobalt carbonyl hydride formed by pumping the volatile products through a $-78°$ (Dry Ice–acetone mixture) trap into a $-196°$ trap. A dark residue will remain in the reactor consisting of unreacted dicobalt octacarbonyl and decomposition products. Unreacted trifluorosilane will be in the $-196°$ trap.

The $-78°$ trap will contain both the trifluorosilyl carbonyl and any carbonyl hydride formed. These are then condensed into a clean 20-ml. glass reactor cooled to $-196°$. About 1.0 g. trifluorosilane is then condensed into the reactor at $-196°$, and the mixture is allowed to stand at room temperature for about 2 hours.

The reactor is then cooled to $-196°$, and any hydrogen formed is removed by pumping. The reactor is allowed to warm to $0°$, and the contents are then pumped through traps cooled to -78 and $-196°$. The trap cooled to $-78°$ will contain the transparent amber crystals of tetracarbonyl(trifluorosilyl)cobalt which melt at $19.4–19.5°$ to form an amber liquid. Yield is 60% based on $Co_2(CO)_8$ employed.

Properties

Tetracarbonyl(trifluorosilyl)cobalt is a white solid which upon vacuum sublimation forms transparent amber crystals which melt at $19.5°$ with decomposition. This compound is kept at $0°$ with an ice bath and sublimed in the vacuum system to minimize decomposition. For example, in manipulating this compound in the vacuum line, distillation traps should be at $0°$ or lower. Gaseous samples, however, are not observed to undergo decomposition at temperatures up to $80°$. The gas-phase infrared spectrum exhibits bands at 2128(m), 2073(s), and 2049(vs) cm.$^{-1}$ due to carbonyl groups[9] and SiF absorptions at 940(s) and 825(s) cm.$^{-1}$.

References

1. A. J. Chalk and J. F. Harrod, *J. Am. Chem. Soc.*, **87**, 1133 (1965).
2. B. J. Aylett and J. M. Campbell, *Chem. Commun.*, **1965**, 217.
3. W. Jetz and W. A. G. Graham, *J. Am. Chem. Soc.*, **89**, 2773 (1967).
4. W. Jetz, P. B. Simons, J. A. J. Thompson, and W. A. G. Graham, *Inorg. Chem.*, **5**, 2217 (1966).
5. A. J. Chalk and J. F. Harrod, *J. Am. Chem. Soc.*, **89**, 1640 (1967).
6. Y. L. Baay and A. G. MacDiarmid, *Inorg. Chem.*, **8**, 986 (1969).
7. T. G. Gibian, and D. S. McKinney, *J. Am. Chem. Soc.*, **73**, 1431 (1951).
8. K. L. Watters, J. N. Brittain, and W. M. Risen, Jr., *Inorg. Chem.*, **8**, 1347 (1969).
9. A. P. Hagen and A. G. MacDiarmid, *ibid.*, **6**, 686 (1967).
10. R. A. Friedel, I. Wender, S. L. Shufler, and H. W. Sternberg, *J. Am. Chem. Soc.*, **77**, 3951 (1955).
11. C. Newman, S. R. Polo, and M. K. Wilson, *Spectrochim. Acta*, **15**, 793 (1959).

14. ISOLEPTIC* ALLYL DERIVATIVES OF VARIOUS METALS

$$C_3H_5X + Mg \rightarrow C_3H_5MgX$$
$$xC_3H_5MgX + MCl_x \rightarrow M(C_3H_5)_x + xMgXCl$$
where M = Si, Ge, Sn, Cr, Ni

Submitted by S. O'BRIEN,† M. FISHWICK,‡ B. McDERMOTT,‡
M. G. H. WALLBRIDGE,‡ and G. A. WRIGHT‡
Checked by G. W. PARSHALL§ and E. R. WONCHOBA§

Many metal allyls and their derivatives may be prepared from an allyl Grignard reagent.[1-4] However, the conditions for the

---

* The term isoleptic is used to indicate that all the ligands attached to the central metal atom are identical in constitution.

† I. C. I. Ltd., Petrochemical and Polymer Laboratory, The Heath, Runcorn, U.K.

‡ Department of Chemistry, The University, Sheffield, U.K.

§ Central Research Department, E. I. du Pont de Nemours & Company, Wilmington, Del. 19898.

various reactions vary depending upon the properties of the particular metal allyl prepared, especially its thermal stability and sensitivity to air. The procedures given below describe a convenient method for the preparation of allylmagnesium chloride and the subsequent preparation of tetraallylsilanes, germanes and stannanes, triallylchromium, and diallylnickel. Although the procedure for the preparation of the Grignard reagent is similar to that reported earlier,[5] it has been modified sufficiently to warrant a detailed description. Allyl bromide may be used to prepare the Grignard reagent instead of allyl chloride, but offers no real advantage over the cheaper chloro derivative. All solvents and reagents are dried and degassed before use.

A. ALLYL MAGNESIUM CHLORIDE

(Allylchloromagnesium)

$$C_3H_5Cl + Mg \rightarrow C_3H_5MgCl$$

Procedure

A dry, 5-l. flask* fitted with a stirrer, a thermometer to monitor the temperature of the reaction mixture, condenser, dropping funnel, and inlet and outlet tubes (the latter being attached to the top of the condenser) for dry nitrogen is charged with 45.5 g. (1.896 moles) of magnesium turnings which have been washed with benzene, followed by acetone, and dried in an oven for 3 hours. After adding 1250 ml. of diethyl ether the system is flushed with dry nitrogen for 30 minutes and a slow stream maintained through the flask for the remainder of the experiment. Small quantities of iodine and ethyl bromide (1–2 ml.) are added to the flask to initiate reaction, and as soon as the reaction has commenced (indicated by loss of the iodine color from the solution), the flask is transferred to an ice-salt bath ($-10°$ to $-15°$) to maintain the temperature of the reaction mixture below $-10°$. A further 1000 ml. of ether is

---

* Smaller quantities of the Grignard reagent may be prepared using proportionate weights of the reagents.

added, followed by 131.7 g. (1.72 moles) of allyl chloride in 900 ml. of ether which is added dropwise with rapid stirring over about 3 hours. Slow addition of the allyl chloride is necessary to prevent the exothermic reaction becoming too vigorous, and to minimize the formation of biallyl (1,5-hexadiene). As the reaction proceeds, the solution usually remains reasonably clear, but darkens to a grey-black color.* The contents of the flask are stirred for a further 30 minutes at −10 to −15°, and the preparation of the Grignard reagent is then complete.

The yield of the Grignard reagent is estimated as follows: A 50 ml. aliquot† is pipetted into a 250-ml. conical flask containing an excess (30 ml.) of 1 M hydrochloric acid. The ether is evaporated, and the excess acid is then back titrated with 1 M sodium hydroxide using methyl red as indicator. Typical yields are of the order of 60% of theoretical.

B. TETRAALLYLSTANNANE

$$4C_3H_5MgCl + SnCl_4 \rightarrow Sn(C_3H_5)_4 + 4MgCl_2$$

The Grignard reagent (127 g., 1.26 moles) is prepared in about $2\frac{1}{2}$ l. of ether and collected in a 5-l. flask. The flask is fitted with a condenser, stirrer, dropping funnel, and an inlet and outlet for nitrogen. After cooling the flask in an ice-salt bath (−10 to −15°), tin tetrachloride (65.6 g., 0.252 mole, 80% of the theoretical amount) is added dropwise with rapid stirring over about one hour. The stirrer is stopped, and the mixture is refluxed under nitrogen for 12 hours. Distilled water (*ca.* 500 ml.) is added slowly to the flask‡ (which is cooled again in an ice-salt bath) to hydrolyze the excess Grignard reagent, fol-

---

* In some experiments a flocculent, white solid may appear as the reaction proceeds; this need not be filtered off, and its appearance does not apparently affect the yield of the Grignard reagent.

† If a white solid is present, the mixture should be stirred and an aliquot of the suspension taken.

‡ An alternative method which avoids the hydrolysis is to filter off the ethereal solution from any solid which has collected in the flask and to remove the ether as described. The yields using this procedure may be slightly lower than the hydrolysis method.

lowed by 500 ml. of 3% hydrochloric acid to dissolve most of the solid deposited on hydrolysis, and the ether layer separated. The water layer is washed twice with ether, and the combined ether extracts dried over calcium chloride overnight. The ethereal solution is filtered from the calcium chloride rapidly in the air. The solution should be tested for chloride at this stage, since an impurity [probably triallyltin chloride (triallylchloro-stannane)] frequently appears in the solution. The chloride may be removed by shaking the ethereal solution with a 5% w/w potassium fluoride solution, and filtering off the white solid fluoride precipitated. The ether layer is then separated and dried over calcium chloride overnight, the ether is distilled off, and the residual liquid is transferred to a 100-ml. Claisen flask. The liquid is then distilled under reduced pressure and the product obtained as a colorless liquid. The yield is 38.4 g. (53.0% theoretical based on addition of tin tetrachloride).

Repeated distillations of the product appear to decrease both the yield and the purity, and always after distillation a yellow involatile oil remains in the distillation flask. The optimum conditions for the distillation are those where the temperature is kept as low as possible; a pressure of 0.2 mm., giving a boiling point of 52°, is suitable.

C. TETRAALLYLSILANE AND TETRAALLYLGERMANE

Both these compounds may be prepared by the action of the appropriate tetrachloride on the Grignard reagent using a procedure identical to that used for the tin compound. The yields of the silicon and germanium compounds are 35 and 45% of theoretical, respectively, based on the metal halide.

Properties

The tetraallyl derivatives of silicon, germanium, and tin are all colorless liquids, with a characteristic odor, which react only

very slowly in air, depositing a white solid.   They do not appear
to possess definite melting points, but freeze to a glass on cooling.
The boiling points, refractive indices, and densities of the compounds are: for silicon, 87°/10 mm.; n_D^{25}, 1.482; d^{20}, 0.837 g./ml.;
germanium, 105°/10 mm.; n_D^{25}, 1.503; d^{20}, 1.015 g./ml.; tin,
52°/0.2 mm.; n_D^{25}, 1.536; d^{20}, 1.179 g./ml., respectively.   The
compounds may be identified by their infrared spectra; all show a
characteristic band near 1630 cm.$^{-1}$ from the olefinic bond, and
their mass spectra, where m/e values correspond to the ions
$M(C_3H_5)_4^+$, $M(C_3H_5)_3^+$, $M(C_3H_5)_2^+$, and $M(C_3H_5)^+$, where
M = Si, Ge, or Sn, may be observed.

D. TRI-h^3-ALLYLCHROMIUM

The reactor flask shown in Fig. 3* is fitted with a stirrer, a
Herschberg funnel—the end of which protrudes past the neck
into *A*—and a combined nitrogen inlet vacuum take-off.   The

* Normal round-bottomed flasks can be used, but the reactor illustrated greatly
facilitates the manipulations with resultant improvement in the product yield.

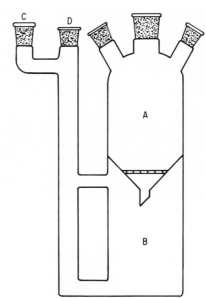

*Fig.   3.   Reactor   for   the
preparation   of   tri-h^3-allyl-
chromium.*

Herschberg funnel is also fitted with a nitrogen inlet and a serum cap. Flasks A and B are separated by a porosity 2 sintered disk. Joint D is stoppered, and C is connected to a nitrogen supply at a slightly positive pressure with respect to that in A. (This is obtained conveniently by using a liquid paraffin head of approximately 6 cm.) With C closed the whole apparatus is evacuated and filled with nitrogen. Anhydrous chromium trichloride (4.7 g., 29.6 mmoles) is quickly placed in A via a powder funnel (the Herschberg funnel being removed and then replaced), and then the whole apparatus including the Herschberg funnel is evacuated and filled with nitrogen a further two times. C is opened, and ether (30 ml.) is syringed into the Herschberg funnel which on addition to A forms a slurry with the chromium trichloride. (The positive pressure in B holds the ether above the sinter.) The reactor is immersed in a trichloroethylene bath maintained at a temperature between -20 and $-30°$ by the addition of solid CO_2, and 9.9 g. (98 mmoles) of the reagent in ether (195 ml., 0.5 M solution) is added over 2 hours with stirring. Stirring is continued at the same temperature for a further 3 hours. After standing overnight at $-78°$, the dark red solution is filtered into B by applying a positive pressure in A. A is washed with ether (two times with 25 ml.) and then quickly fitted with three stoppers. The flask B is connected through D, via an inverted U tube, to a nitrogen-purged, round-bottomed flask connected to a vacuum line. This flask is cooled in liquid nitrogen, and the solvent is removed under reduced pressure (0.1–0.5 mm. Hg) at $-40°$.

After removal of all the ether the apparatus is filled with nitrogen, one of the stoppers in A is replaced by a serum cap, and pentane (150 ml.) is syringed into A and precooled for 10 minutes before it is added to B. A dip tube is inserted through D, the other end of which is connected, using silicone tubing, to a second reactor similar to the one illustrated in Fig. 3. The dip tube is long enough to reach the bottom of flask B and is, conveniently, a glass tube which can be moved up and down through a cone adaptor fitted with a screw cap. The seal between the adaptor and glass tube is maintained by silicone rubber and Teflon washers which can be tightened by the cap. Before inserting the dip tube through D, the second reactor and connecting tube (the end of which is plugged) are evacuated and filled with nitrogen two times. With the dip tube inserted to the bottom of B a positive pressure is applied at C and the solution pushed as quickly as possible into the second reactor and filtered. The initial reactor is extracted with pentane (two times with 50 ml.) which is also transferred to the second reactor and filtered. The yield (obtained by analyzing an aliquot of solution for chromium) is 3.6–4.1 g. (69–79% theoretical). The product can be further purified by concentrating the pentane solution to approximately two-thirds volume at $-40°$ and allowing it to crystallize at $-78°$. Tri-h^3-allylchromium and its thermal decomposition products are pyrophoric in air.

E. DI-h^3-ALLYLNICKEL

This compound can be prepared similarly to tri-h^3-allylchromium in 80% yield starting from anhydrous nickel dichloride. However, owing to the volatility of di-h^3-allylnickel, which co-distills with ether very readily, the solvent is best removed at $-78°$ and a pressure of 10^{-2} mm. Hg. The product can be obtained as pale yellow crystals from pentane.

The transition-metal allyl complexes are air- and temperature-sensitive solids: $Cr(allyl)_3$, m.p. *ca.* $70°$; $Ni(allyl)_2$, m.p. *ca.* $0°$. The infrared spectrum of both compounds indicates that the bonding of the allyl group to the metal involves π electrons (the olefinic bond appearing at 1520 and 1493 cm.$^{-1}$, respectively); they can be identified by their mass spectra.

References

1. G. Wilke, *et al., Angew. Chem.* (Intern. Ed.), **5**, 151 (1966).
2. M. L. H. Green and P. L. I. Nagy, *Advan. Organometall. Chem.*, **2**, 325 (1964).
3. J. K. Becconsall, B. E. Job, and S. O'Brien, *J. Chem. Soc.* (A), **1967**, 423.
4. E. Kurras and P. Klimsch, *Deut. Akad. Wiss. Monat.*, **6**, 735 (1964).
5. O. Grummitt, E. P. Budewitz, and C. C. Chudd, *Org. Syn.*, coll. vol. 4, 749 (1963).

Chapter Three

COMPOUNDS CONTAINING
METAL-TO-METAL BONDS

15. RHENIUM AND MOLYBDENUM COMPOUNDS CONTAINING QUADRUPLE BONDS

Submitted by ALICIA B. BRIGNOLE* and F. A. COTTON*

The occurrence of a quadruple bond was first recognized[1] in the case of $[Re_2Cl_8]^{2-}$. It has since been shown that the quadruply bonded Re_2 entity will persist through a variety of ligand substitution reactions,[2] and an extensive chemistry of this quadruply bonded diatomic unit has already been developed. In general, the easiest access to this class of compounds is via an $[Re_2Cl_8]^{2-}$ salt. A quick and convenient method for the preparation of this species and its isolation as the tetra-n-butylammonium salt is therefore described. The preparation of the bromo analog $[Re_2Br_8]^{2-}$ is also given.

Another important and extensive class of dirhenium derivatives are the carboxylate-bridged molecules,[2,3] $Re_2(O_2CR)_4X_2$; general procedures for obtaining these compounds from $[Re_2Cl_8]^{2-}$ are thus of value and are presented here.

* Massachusetts Institute of Technology, Cambridge, Mass. 02139.

Dinuclear molybdenum(II) molecules[4,5] also contain quadruple bonds.[5] The acetate $Mo_2(O_2CCH_3)_4$ is also a very useful starting material for the preparation of other complexes of lower-valent molybdenum.[5-7] The preparations of $Mo_2(O_2CCH_3)_4$ and $Mo_2(O_2CC_6H_5)_4$ are therefore described.

References

1. F. A. Cotton, *Inorg. Chem.*, **4**, 334 (1965); F. A. Cotton and C. B. Harris, *ibid.*, **6**, 924 (1967).
2. See also M. J. Bennett, W. K. Bratton, F. A. Cotton, and W. R. Robinson, *ibid.*, **7**, 1570 (1968), and earlier papers cited therein.
3. F. A. Cotton, C. Oldham, and W. R. Robinson, *ibid.*, **5**, 1798 (1966).
4. D. Lawton and R. Mason, *J. Am. Chem. Soc.*, **87**, 921 (1965).
5. J. Brencic and F. A. Cotton, *Inorg. Chem.*, **8**, 7 (1969).
6. I. R. Anderson and J. C. Sheldon, *Australian J. Chem.*, **18**, 271 (1965).
7. J. Brencic and F. A. Cotton, *Inorg. Chem.*, **8**, 1060 (1969).

A. OCTAHALODIRHENATE(III) ANIONS

$$Re_3Cl_9 \xrightarrow{\text{molten } (C_2H_5)_2NH_2Cl} Re_2Cl_8{}^{2-}$$
$$Re_2Cl_8{}^{2-} + 8Br^- \rightarrow Re_2Br_8{}^{2-} + 8Cl^-$$

Checked by Z. DORI*

The octachlorodirhenate(III) anion, $[Re_2Cl_8]^{2-}$ has been prepared in several ways, including (*a*) reduction of perrhenate ion in acid solution with molecular hydrogen at high temperature and pressure;[1] (*b*) reduction of perrhenate ion with hypophosphorus acid, in a solution of constant boiling hydrochloric acid;[1,2] (*c*) displacement of the carboxylate ligands in Re_2-$(OOCR)_4Cl_2$ with chloride ion in concentrated hydrochloric acid;[1,2] (*d*) reaction of trirhenium nonachloride in molten diethylammonium chloride.[3]

The first method gives yields of 30–40%, but corrosion of the

* Temple University, Philadelphia, Pa. 19122.

pressure bomb and other practical difficulties make it relatively undesirable. Method *b* is convenient, but the yields never exceed ∼40%. Method *c* is of no practical interest, since the most efficient way to prepare the $Re_2(O_2CR)_4Cl_2$ compounds themselves is by reaction of RCOOH with $[Re_2Cl_8]^{2-}$, as described later. Method *d* is simple; it uses a commercially available starting material and gives yields of up to 65%.

Procedure

(a) *The* $[Re_2Cl_8]^{2-}$ *Ion.* A thoroughly ground mixture of 2 g. of trirhenium nonachloride* and 30 g. of diethylammonium chloride is placed in a glass tube of *ca.* 2 cm. i.d. and 20 cm. long, equipped with outlet and inlet tubes, each bearing a stopcock near the top to permit a nitrogen flow during the reaction.

It is important that the preparation be run entirely in an inert atmosphere. This is accomplished by evacuating the system and refilling it with dry prepurified nitrogen. This procedure should be repeated at least three times to ensure complete absence of oxygen.

The mixture is fused under a nitrogen flow and held just above the melting point of the amine hydrochloride (217–220°) by means of a metal bath. The temperature should not be permitted to rise more than a few degrees above the melting point of the amine hydrochloride or the yield will be significantly lowered. The color quickly changes from pink to dark green, and the heating is continued for 5–10 minutes after the solid is completely melted. The reaction vessel is removed from the metal bath, and the melt is allowed to solidify. After it has reached room temperature, it may be handled in air. Approximately 150 ml. of 6 *N* HCl is used to dissolve the green solid. Portions of acid are added to the reaction vessel, and the solid is pried loose from the tube and chopped up using a ceramic

---

* Available from Shattuck Chemical Co., Boulder, Colo., or cf. *Inorganic Syntheses*, **12**, 193 (1970).

spatula. The solution is then filtered through a fritted glass disk; a brown solid (of as yet unknown constitution) is separated on filtration. The octachlorodirhenate($2-$) salt can be precipitated from the green solution by addition of the chloride of a large cation. A yield of 60–65% (2.4 g.) is obtained using tetrabutylammonium chloride as a precipitant. The blue-green solid is washed with 10 ml. of hydrochloric acid, 10 ml. of ethanol, and 10 ml. of ether and then dried in vacuum over anhydrous copper sulfate. *Anal.* Calcd. for $[Re_2Cl_8][(C_4H_9)_4N]_2$: C, 33.60; H, 6.35; N, 2.46. Found: C, 33.51; H, 6.39; N, 2.36.

■ *Note.* *The use of a wider tube and larger amounts of reactants decreases the percentage yield.*

(*b*) *The Octabromodirhenate(2−) Anion.* This is obtained from $[Re_2Cl_8]^{2-}$ by halogen exchange.

To a solution of 1.5 g. of $[(C_4H_9)_4N]_2[Re_2Cl_8]$ in 200 ml. of methanol is added 50 ml. of 48% aqueous hydrogen bromide. Evaporation of the solvent at *ca.* 60° yields olive-green crystals of $[(C_4H_9)_4N]_2[Re_2Br_8]$, which are filtered and washed with ethanol and ether. Yield is 98%.

Properties

The $[Re_2X_8]^{2-}$ ions are stable in acid solutions. The tetrabutylammonium salts are soluble in methanol, acetone, acetonitrile, and various other solvents, giving blue (Cl) or green-blue (Br) solutions which are stable in air in presence of a few drops of concentrated hydrochloric or hydrobromic acid. They can be recrystallized from methanol.

References

1. F. A. Cotton, N. F. Curtis, B. F. G. Johnson, and W. R. Robinson, *Inorg. Chem.,* **4,** 326 (1965).
2. F. A. Cotton, N. F. Curtis, and W. R. Robinson, *ibid.,* **4,** 1696 (1965).
3. R. Bailey and J. McIntyre, *ibid.,* **5,** 1940 (1966).

B. TETRAKIS(CARBOXYLATO)DIHALODIRHENIUM(III) COMPOUNDS

$$[Re_2X_8]^{2-} + 4RCOOH \rightarrow Re_2(O_2CR)_4X_2 + 4HX + 2X^-$$
$$Re_2(O_2CCH_3)_4X_2 + 4ArCOOH \rightarrow Re_2(O_2CAr)_4X_2 + 4CH_3COOH$$

Checked by Z. DORI*

Compounds of the type $Re_2(O_2CR)_4Cl_2$ were first reported by Taha and Wilkinson[1] who obtained them in low yield by reaction of nonachlorotrirhenium with a carboxylic acid. They are also formed on prolonged heating of *trans*-$ReOCl_3[P(C_6H_5)_3]_2$ with carboxylic acids.[2] On heating octahalogenodirhenium(2−) salts with carboxylic acids nearly quantitative conversions to $Re_2(O_2CR)_4X_2$ may be quickly effected.[3] However, when the R group is aryl, it is expeditious to use a two-step process in which the acetato compound is first prepared and then exchanged with the desired aryl acid.[3]

Procedure

(a) *Preparation of* $Re_2(O_2CCH_3)_4Cl_2$. A solution of 1.0 g. of $[(C_4H_9)_4N]_2[Re_2Cl_8]$ in 40 ml. of glacial acetic acid and 10 ml. of acetic anhydride is boiled under reflux for 2 hours, with a slow stream of nitrogen passing through the solution. The solution becomes brown and crystals of the acetate are deposited as the reaction proceeds. The *cool* solution is filtered in air to separate the crystalline product, which is washed with ethanol, then ether, and dried in a vacuum. Beginning with $[(C_4H_9)_4N]_2$-$[Re_2Br_8]$, the tetrakis(acetato)dibromodirhenium may be made in the same way. Yields are *ca.* 95%. *Anal.* Calcd. for $Re_2(O_2$-$CCH_3)_4Cl_2$: C, 14.1; H, 1.78; Cl, 10.4. Found: C, 13.9; H, 1.75; Cl, 10.4. Calcd. for $Re_2(O_2CCH_3)_4Br_2$: C, 12.5; H, 1.57; Br, 20.8. Found: C, 12.4; H, 1.51; Br, 20.6.

(b) *Preparation of Arylcarboxylato Compounds.* A mixture of 1.0 g. of $Re_2(O_2CCH_3)_4X_2$ and about 5 g. (a large excess) of the

* Temple University, Philadelphia, Pa. 19122.

arylcarboxylic acid is placed in a glass vessel which is equipped with outlet and inlet tubes near the top to permit a nitrogen flow during the reaction. The system is evacuated and filled with nitrogen several times to remove oxygen and thereafter a slow nitrogen flow is maintained. The mixture is fused using an oil or metal bath and held at 200° or a little above the melting point of the acid, whichever is higher. The color of the melt changes from orange to red or brown as the reaction proceeds. Reaction time varies from 2 to 5 hours depending on the acid; the reaction may be assumed complete when no further change or intensification of color is observed, but the yield is increased if the heating is continued for 2 hours after the time when no more acetic acid is evolved. This can be judged by the odor of the effluent nitrogen or by using moist pH test paper. For those products which are soluble in chloroform, completeness of reaction is not critical, but for those which are not, it is important as there is no convenient method of subsequent purification.

TABLE I

$Re_2(O_2CR)_4X_2$		Analyses						Soluble in Cl_3CH	Color
		Calculated			Found				
R	X	%C	%H	%Cl or Br	%C	%H	%Cl or Br		
C_6H_5	Cl	36.0	2.17	7.68	36.3	2.01	7.55	Yes	Red
C_6H_5	Br	33.1	1.98	15.7	33.0	1.93	16.6	Yes	Red
p-$C_6H_4CH_3$	Cl	39.1	2.84	7.89	39.2	2.79	7.78	Yes	Red
$1,3,5$-$C_6H_2(CH_3)_3$	Cl	43.8	4.03	6.46	43.8	4.12	6.53	Yes	Pink-red
m-$C_6H_4CH_3$	Cl	38.8	2.78	7.98	38.9	2.72	7.55	Yes	Red
p-$C_6H_4OCH_3$	Cl	36.7	2.69	6.77	36.6	2.75	6.83	Yes	Brown
p-$C_6H_4OCH_3$	Br	33.8	2.49	14.1	33.7	2.37	12.0	Yes	Brown
p-$C_6H_4NH_2$	Cl	34.1	2.45	7.17	34.2	2.58	7.37	No	Brown
p-C_6H_4Cl	Cl	31.6	1.51	20.0	32.2	1.74	20.7	No	Red
p-C_6H_4Cl	Br	29.1	1.39	12.3	30.1	1.40	12.8	No	Red
p-C_6H_4Br	Cl	27.1	1.29	25.7	27.7	1.39	26.2	No	Red
p-C_6H_4Br	Br	25.2	1.21	35.9	25.9	1.31	36.4	No	Red

The melt is allowed to cool. It may then be handled in air. The excess of acid is extracted with diethyl ether. For those products which are soluble in chloroform (see Table I) recrystallization is carried out using this solvent and the product dried in vacuum at 100° for 12 hours. The yields vary from 70 to 90% based on $Re_2(O_2CCH_3)_4X_2$.

Properties

The orange $Re_2(O_2CCH_3)_4Cl_2$ and brown $Re_2(O_2CCH_3)_4Br_2$ are air-stable and practically insoluble in common solvents. The properties of some arylcarboxylato compounds are summarized in Table I.

The structures of these molecules, first postulated by Taha and Wilkinson,[1] with bridging RCOO groups and the ligands X lying at each end to give a linear XReReX chain have been confirmed in the case of $Re_2(O_2CC_6H_5)_4Cl_2$ by a recent x-ray study.[4]

References

1. F. Taha and G. Wilkinson, *J. Chem. Soc.*, **1963**, 5406.
2. G. Rouschias and G. Wilkinson, *ibid.* (*A*), **1966**, 465.
3. F. A. Cotton, C. Oldham, and W. R. Robinson, *Inorg. Chem.*, **5**, 1798 (1966).
4. M. J. Bennett, W. K. Bratton, F. A. Cotton, and W. R. Robinson, *ibid.*, **7**, 1570 (1968).

C. TETRAKIS(CARBOXYLATO)DIMOLYBDENUM(II) COMPOUNDS

$$2Mo(CO)_6 + 4RCOOH \rightarrow Mo_2(O_2CR)_4$$

Checked by Z. DORI* and G. WILKINSON†

These compounds have been obtained by only one general route, viz., by the interaction of carboxylic acids with molybde-

* Temple University, Philadelphia, Pa. 19122.
† Imperial College of Science and Technology, London, S.W. 7, England.

num hexacarbonyl.[1-3] Other products of the reaction have not been fully elucidated. The importance of carrying out all reactions strictly under nitrogen cannot be overemphasized. Yields suffer drastically if this is not done effectively.

Procedure

(*a*) *Tetrakis(acetato)dimolybdenum.* A solution of 2 g. of molybdenum hexacarbonyl in *ca.* 100 ml. of glacial acetic acid and 10 ml. of acetic anhydride is refluxed at 140–150°, under nitrogen. At the beginning the solution is yellow, but after 12 hours yellow needles separate from the solution which becomes brownish. After approximately 20 hours the solution is cooled and the crystalline product is filtered in air, washed with ethanol and ether, and dried in vacuum. Yield is 1.5 g. [37% based on $Mo(CO)_6$]. *Anal.* Calcd. for $C_8H_{12}O_8Mo_2$: C, 22.4; H, 2.83; Mo, 44.8.   Found: C, 22.3; H, 2.68; Mo, 45.2.

(*b*) *Tetrakis(arylcarboxylato)dimolybdenum Compounds.* Molybdenum hexacarbonyl 4 g.) and 5 g. of acid are added to 150 ml. of bis(2-methoxyethyl) ether. Part of the carbonyl is insoluble at room temperature. The diglyme should be dried over molecular sieves and freshly distilled, and the reaction is carried out under nitrogen. The importance of (*a*) *thoroughly* cleaning up the solvent and (*b*) *strictly* excluding oxygen cannot be overemphasized. Yields suffer drastically when these precautions are neglected. The solution is brought to reflux at 150°. During the refluxing, unreacted carbonyl sublimes from the reaction mixture into the condenser and from time to time must be pushed back into the reaction flask. After about 15 minutes the solution becomes yellow, and after about an hour a yellow to orange solid begins to separate from the reaction mixture, leaving a brownish supernatant liquid. After 4 hours the reaction mixture is cooled to 25°. The orange-yellow crystals are collected, washed with anhydrous ether, and dried in vacuum at 70° for 24 hours.   All these operations are carried out under nitrogen.   Yields up to 65% can be obtained. Analyses of representative products are given in Table II.

TABLE II

Compound	Calculated				Found				Color
	% C	% H	% Mo	% Cl	% C	% H	% Mo	% Cl	
$Mo_2(OOCC_6H_5)_4$	49.6	2.98	29.3	....	49.4	2.9	28.6	....	Yellow
$Mo_2(OOCC_6H_4pCl)_4$	41.4	1.98	....	17.4	41.3	2.01	....	17.1	Orange
$Mo_2(OOCC_6H_4pCH_3)_4$	52.2	3.86	26.2	....	52.4	4.21	25.2	....	Orange
$Mo_2(OOCC_6H_4pOCH_3)_4$	48.2	3.54	24.05	....	48.5	3.68	24.01	....	Yellow

Properties

The compounds are all yellow to orange in color.   They vary in stability and sensitivity to air.   The acetate is a yellow solid which is practically insoluble in all common solvents.   It slowly decomposes (turning green and finally dark blue) over a period of days to weeks at room temperature.   Decomposition occurs even in an inert atmosphere and does not appear to be appreciably faster in dry air.

The arylcarboxylato compounds are more soluble, e.g., in $CHCl_3$ and acetone, but the solutions are extremely air-sensitive. Except for the benzoate they are all very air-sensitive even as solids and must be handled entirely under nitrogen.

References

1. T. A. Stephenson, E. Bannister, and G. Wilkinson, *J. Chem. Soc.*, **1964**, 2538.
2. E. Bannister and G. Wilkinson, *Chem. Ind.*, **1960**, 319.
3. E. W. Abel, A. Singh, and G. Wilkinson, *J. Chem. Soc.*, **1959**, 3097.

16. TETRAKIS(ACETATO)DIRHODIUM(II) AND SIMILAR CARBOXYLATO COMPOUNDS

$$\text{RhCl}_3{\cdot}3\text{H}_2\text{O} \xrightarrow{\text{CH}_3\text{COONa/CH}_3\text{COOH}} [\text{Rh(OCOCH}_3)_2]_2$$

Submitted by G. A. REMPEL,* P. LEGZDINS,* H. SMITH,* and G. WILKINSON*
Checked by D. A. UCKO†

The dimeric tetraacetato bridged $\text{Rh}_2(\text{OCOCH}_3)_4$ has been obtained by the interaction of ammonium chlororhodate(III) or rhodium(III) hydroxide with acetic acid.[1-3] Other (carboxylato)rhodium(II) compounds were prepared directly in a similar way or from the acetate by exchange.[2,3] Halo carboxylates (RCOO^-, $\text{R} = \text{CCl}_3$, CF_3, CH_2Cl, etc.) were prepared also by interaction of rhodium trichloride with the appropriate sodium salt in ethanol.[4] The carboxylates are normally first isolated as a solvent adduct, e.g., $[\text{Rh(OCOR)}_2{\cdot}\text{C}_2\text{H}_5\text{OH}]_2$ but are easily converted to the unsolvated complex. The acetate is readily prepared in a modification of this last procedure. A similar method is satisfactory for the preparation of other lower carboxylates as well as halo carboxylates.

Procedure

Commercial rhodium trichloride trihydrate (5.0 g.) and sodium acetate trihydrate (10.0 g.) in glacial acetic acid (100 ml.) and absolute ethanol (100 ml.) were gently refluxed under nitrogen for an hour.

The initial red solution rapidly becomes green, and a green solid is deposited. After cooling to room temperature the green

---

* Inorganic Chemistry Laboratories, Imperial College, London, S.W. 7, England.
† Massachusetts Institute of Technology, Cambridge, Mass. 02139.

solid is collected by filtration through a Büchner or sintered filter funnel.

The crude product is dissolved in boiling methanol (*ca.* 600 ml.) and filtered; after concentration to about 400 ml. the solution is kept in a refrigerator overnight. After collection of the crystals, the solution is concentrated and cooled to yield a further small amount of the methanol adduct $[Rh(OCOCH_3)_2]_2 \cdot 2CH_3OH$.

The blue-green adduct was heated in vacuum at 45° for 20 hours to yield emerald-green crystals of $[Rh(OCOCH_3)_2]_2$. A check on the removal of methanol can be made periodically by taking an infrared spectrum. Overall yield is 3.2 g. (76% based on $RhCl_3 \cdot 3H_2O$). *Anal.* Calcd.: C, 21.74; H, 2.74. Found: C, 21.79; H, 2.99.

Properties

The copper-acetate-type structure has been shown by x-ray diffraction.[5, 6] The complex is diamagnetic; it is only slightly soluble in water, methanol, acetone, etc., giving green solutions. Adducts with a variety of ligands have been characterized.[2-4]

The infrared spectrum has bands at 1580(s), 1425(s), and 1355(m) in Nujol mulls [in hexachlorobutadiene 1445(s), 1415(s) and 1350(m)] due to carboxylate frequencies, as well as CH_3 absorption.

References

1. L. A. Nazarova, I. I. Cherniaov, and A. S. Morozova, *Zh. Neorgan. Khim.* **10**, 291 (1965).
2. S. A. Johnson, H. R. Hunt, and H. M. Neuman, *Inorg. Chem.*, **2**, 950 (1963).
3. T. A. Stephenson, S. M. Morehouse, A. R. Powell, P. J. Heffer, and G. Wilkinson, *J. Chem. Soc.*, **1965**, 3632.
4. G. Winkhaus and P. Ziegler, *Z. Anorg. Allgem. Chem.*, **350**, 51 (1967).
5. M. A. Porai-Koshits and A. S. Antsyshkina, *Dokl. Akad. Nauk. SSR*, **146**, 1102 (1962).
6. F. A. Cotton, B. G. De Boer, M. D. LaPrade, J. R. Pipal, and D. A. Ucko, *J. Am. Chem. Soc.*, **90**, 2926 (1970).

17. DODECACARBONYLTRIRUTHENIUM AND -TRIOSMIUM

Submitted by BRIAN F. G. JOHNSON* and JACK LEWIS*
Checked by C. W. BRADFORD†

Dodecacarbonyltriruthenium[1-4] and -triosmium[5] have been known for many years. Originally formulated as dimetal nonacarbonyls, $M_2(CO)_9$, analogous to $Fe_2(CO)_9$, their molecular configurations have recently been determined by x-ray diffraction studies and the trimeric formulation $M_3(CO)_{12}$ (M = Ru[6,7] or Os[7]) established. Although formally analogous to $Fe_3(CO)_{12}$[8] these compounds have a different structure which consists of a triangular arrangement of three metal atoms with four terminal carbonyl groups per metal.

A. DODECACARBONYLTRIRUTHENIUM

$$Ru(C_5H_7O_2)_3 + H_2 + CO \rightarrow Ru_3(CO)_{12} \text{ (+ reduction products)}$$

The early preparations gave poor yields but highly efficient methods have been developed recently. Optimum yields are obtained by the method of Pino and his coworkers[9] in which tris(2,4-pentanedionato)ruthenium(III) is treated with equimolar mixtures of hydrogen and carbon monoxide at moderate temperatures and pressures (140–160°, 200–300 atmospheres). However, this method is limited by the availability of the tris-(2,4-pentanedionato)ruthenium(III) which is obtained in only low yields from the readily available ruthenium trichloride hydrate. The method given here is a modification on the Pino method.

* Department of Chemistry, University College London, Gower Street, London, WC 1, England.
† Johnson Matthey and Co., Ltd., Exhibition Grounds, Wembley, Middlesex, England.

Reactions are carried out in a 1-l. autoclave capable of withstanding pressures up to 300 atmospheres and operating at maximum temperatures of 300°. All solvents must be freshly distilled before use and all manipulations carried out in a well-ventilated hood. It is essential that the sodium acetylacetonate, (2,4-pentanedionato)sodium, be freshly prepared before use.*

Ruthenium trichloride hydrate (5 g.), sodium acetylacetonate (7 g.), and methyl alcohol (140 ml.) are placed in the autoclave *in that order*. Hydrogen (40 atmospheres) and carbon monoxide (120 atmospheres) (i.e., total initial pressures = 160 atmospheres at room temperature) are then added and the reaction mixture heated at 165° for 4 hours. When cold the pressure is released and the crude orange crystalline dodecacarbonyltriruthenium separated by filtration. The mother liquor is evaporated to dryness and any additional product extracted into hot hexane in a Soxhlet apparatus. The combined products are then recrystallized from hot hexane.† Yields vary slightly from preparation to preparation but are usually in the range 50–55% (2.5 g.). (The checker obtained a yield of 3.0 g., 70%.)

B. DODECACARBONYLTRIOSMIUM

$$3OsO_4 + 24CO \rightarrow Os_3(CO)_{12} + 12CO_2$$

Dodecacarbonyltriosmium was first obtained from the reaction of osmium(VIII) tetraoxide with carbon monoxide in the gas phase.[5] Yields by this method are low. More recently, Bradford and Nyholm[10] have obtained this carbonyl in high yields from the reaction of osmium tetraoxide with carbon monoxide in xylene under conditions of moderate pressure (128 atmospheres) and temperature (175°). The method

---

* The checker reports the following procedure: 2.3 g. NaOH as a 40% w/v aqueous solution was added to 5.74 g. acetylacetone (2,4-pentanedione) with stirring. The resultant white, virtually solid, mass was cooled and added to the ruthenium trichloride.

† The checker reports that light petroleum (60–80°) can also be used and that a better-looking product was obtained by a further recrystallization from acetone.

described here involves the carbonylation of osmium tetraoxide in alcohols. By this method yields of up to 85% have been obtained. Other minor products from this reaction are the trinuclear derivatives $HOs_3(CO)_{10}(OMe)$ and $Os_3(CO)_{10}(OMe)_2$.[11]

The reaction is carried out in a 100-ml. rocking autoclave* capable of withstanding pressures up to 300 atmospheres and operating at maximum temperatures of 300°. *Owing to the toxicity of osmium compounds, all manipulations must be carried out in a well-ventilated hood.*

Osmium tetraoxide (2 g.) and methyl alcohol (30 ml.) are treated with carbon monoxide (75 atmospheres) at 125° for 12 hours. On cooling, the yellow crystals of dodecacarbonyltriosmium, $Os_3(CO)_{12}$, are separated by filtration, washed with acetone, and dried *in vacuo* (20°, 10^{-1} mm. Hg). The sample thus obtained is usually sufficiently pure for most purposes. Further purification may be carried out either by sublimation at 130° *in vacuo* (10^{-2} mm. Hg) or by recrystallization from hot benzene. The yield of the decacarbonyl is usually in the range 70–80% (1.63 g.).

Properties

Dodecacarbonyltriruthenium is an orange crystalline solid, soluble in a wide variety of organic solvents and volatile. Dodecacarbonyltriosmium is yellow, shows only limited solu-

---

* Larger autoclaves containing a 100-ml. glass container may be employed as an alternative. The checker reports satisfactory results with a nonrocking autoclave.

TABLE I **Infrared Spectra of $Ru_3(CO)_{12}$ and $Os_3(CO)_{12}$**
  (ν_{CO}) **cm.$^{-1}$ (n-hexane)***

$Ru_3(CO)_{12}$	2060(s), 2030(s), 2010(m)
$Os_3(CO)_{12}$	2070(s), 2036(s), 2015(m), 2003(m)

* J. Knight and M. J. Mays, *Chem. Commun.*, **1969**, 384.

bility in organic solvents, and is volatile. The infrared spectra of these carbonyls are similar but differ from that of $Fe_3(CO)_{12}$ (Table I) in that no bridging CO bands are observed in either the solid state or in solution. These spectra are useful for identification and for checking purity.

References

1. L. Mond, N. Hirtz, and M. D. Cowap, *Proc. Chem. Soc.*, **26**, 67 (1910); *J. Chem. Soc.*, **97**, 798 (1910); *Z. Anorg. Allgem. Chem.*, **68**, 207 (1910).
2. L. Mond and A. E. Wallis, *J. Chem. Soc.*, **121**, 29 (1922).
3. W. Manchot and W. J. Manchot, *Z. Anorg. Allgem. Chem.*, **226**, 385 (1936).
4. W. Hieber and H. Fischer, DRP 695589 (1940); *C. A.*, **35**, 5657 (1941).
5. W. Hieber and H. Stallman, *Z. Electrochem.*, **49**, 288 (1943).
6. R. Mason and A. I. M. Rae, *J. Chem. Soc.*, **1968**, 778.
7. E. R. Corey and L. F. Dahl, *J. Am. Chem. Soc.*, **83**, 2203 (1961); *Inorg. Chem.*, **1**, 521 (1962).
8. C. H. Wei and L. F. Dahl, *J. Am. Chem. Soc.*, **88**, 1821 (1966).
9. G. Braca, G. Sbrana, and P. Pino, *Chim. Ind. (Milan)*, **46**, 206 (1964); DRP 1216276 (1966); *C. A.*, **65**, 8409 (1966).
10. C. W. Bradford and R. S. Nyholm, *Chem. Commun.*, **1967**, 384.
11. B. F. G. Johnson, P. A. Kilty, and J. Lewis, *J. Chem. Soc.* (A), **1968,** 2859

18. DODECACARBONYLTETRAIRIDIUM

$$Na_3IrCl_6 + CO \rightarrow Ir_4(CO)_{12}$$

Submitted by L. MALATESTA,* G. CAGLIO,* and M. ANGOLETTA*
Checked by E. M. KAISER† and M. F. FARONA‡

Dodecacarbonyltetrairidium, $Ir_4(CO)_{12}$, can be prepared in good yields by the method used for the first time by Hieber and Lagally[1] from iridium chloride or from an alkali hexachloro-

* Centro di studio per la sintesi e la struttura dei composti dei metalli di transizione nei bassi stati di ossidazione del C.N.R. and Istituto di Chimica Generale dell'Universita, Via G. Venezian 21, 20133 Milan, Italy.
† The Ohio State University, Columbus, Ohio 43210.
‡ The University of Akron, Akron, Ohio 44304.

iridate(III) or (IV) by the action of carbon monoxide at high pressure (250 atmospheres) and moderate temperature (100–150°). In very low yield it can be obtained from hydrated iridium trichloride in a stream of carbon monoxide at atmospheric pressure at 150°.[2] The recent method of Chaston and Stone,[3] by which $Ir_4(CO)_{12}$ can be obtained from a methanol solution of the so-called "soluble iridium(III) chloride" and carbon monoxide at 50–60 atmospheres, represents an improvement in comparison to the previous method but still requires a pressure vessel. Also the preparation of the soluble iridium chloride is quite time-consuming.

The following procedure by which sodium hexachloroiridate(III), Na_3IrCl_6, is transformed directly, under appropriate conditions, into dodecacarbonyltetrairidium, $Ir_4(CO)_{12}$, has many advantages. It uses as starting material the readily available sodium hexachloroiridate(IV), Na_2IrCl_6. It gives rather high yields, and it does not require high pressure or special equipment.

Procedure

■ *Caution.   The reaction must be carried out in a well-ventilated hood.*

The quantity of sodium hexachloroiridate(IV), obtained from the complete consumption of 1.7 g. (8.85 mmoles) iridium metal in the manner described elsewhere,[4] or 5.0 g. of the commercial hexahydrate is dissolved in 100 ml. of ethanol and the solution immediately filtered.   A solution of 1.4 g. (9.4 mmoles) of sodium iodide in 50 ml. ethanol is then added to the solution. The fast reduction of the iridate(IV) ion, $[IrCl_6]^{2-}$, to iridate-(III), $[IrCl_6]^{3-}$, by iodide ions is a well-known reaction;[5] when it is carried out under these conditions, the sodium hexachloroiridate(III) precipitates almost quantitatively and can easily be filtered, washed with ethanol, and dried at 100°.

The hexachloroiridate(III)* (*ca.* 4.8 g.) is now transferred to a 250-ml., three-necked flask, and 7.9 g. (53 mmoles) sodium iodide, 100 ml. of methanol, and 5 ml. of water are added. The flask is equipped with a reflux condenser, a magnetic stirring bar is introduced, and one side neck is attached to a source of carbon monoxide. The gas leaves the flask through the condenser which is connected to a bubble counter. The other side neck is stoppered. The flask is first flushed with carbon monoxide, with vigorous stirring, and then the gas stream is reduced so that about one bubble every 2 seconds passes through the bubble counter. The flask is then gently heated on an electric heating mantle, so that the methanol refluxes very slowly.

In about 4 hours all the insoluble hexachloroiridate(III) is transformed into a soluble compound† and a red-brown solution is obtained. The heating is not suspended, and the current of CO is increased from one bubble every 2 seconds to about two bubbles every 1 second and maintained at that rate until the solution is cold. Now 2.9 g. (17.7 mmoles) of finely ground potassium carbonate‡ is added, and the solution is vigorously stirred at room temperature for about 40 hours, maintaining the current of CO, and then it is filtered in an inert atmosphere (nitrogen or carbon monoxide). The filtrate is set aside under nitrogen.

The filtered solid, mainly $Ir_4(CO)_{12}$ with some water-soluble

---

* The product so obtained may contain some sodium chloride and water; the purity can be checked by weighing about 100 mg. of the solid in a platinum crucible and heating the crucible in an oven at about 1400° to constant weight. The residue is pure iridium.

† The exact nature of this compound is complex and not known with certainty. It may be a carbonyliodoiridate such as $Na[IrCOI_5]$, $Na[Ir(CO)_2I_4]$, or $Na[Ir(CO)_2I_2]$, or a mixture of the three. The essential fact is that it is soluble in ethanol and it is not affected by solid potassium carbonate at room temperature.

‡ Without addition of K_2CO_3, reaction proceeds slowly to give $[Ir(CO)_2I_2]^-$, but no carbonyl is obtained. In the presence of K_2CO_3, $[Ir_4(CO)_{11}H]^-$ is probably formed in the early stages. This is then slowly converted into $Ir_4(CO)_{12}$, when the basicity of the solution is lowered by the reaction of CO with hydroxide ion: $CO + OH^- + K^+ \rightarrow HCOOK$.

impurities, is washed four times with 2 ml. water, then twice with 10 ml. ethanol, and finally with hexane and dried at 50°. The yields vary from 1.8 to 1.9 g. (74–78%) with respect to the original iridium. The solution contains a fraction of the original iridium as soluble alkali pentadecacarbonylhexairidate [$Ir_6(CO)_{15}]^{2-}$, which can be recovered as the tetraethylammonium salt by adding 500 mg. of tetraethylammonium chloride to the solution and treating it with about the same volume of water. The brown crystalline solid (about 0.2 g.) is filtered in a nitrogen atmosphere, washed with water, and dried. It must be kept in inert atmosphere or in vacuum.

Properties

Dodecacarbonyltetrairidium can be very slowly sublimed in a carbon monoxide stream at about 120°.* Sublimed carbonyl melts at 210° if rapidly heated, but when heated slowly it begins to decompose at about 170°. The melting point is therefore not very significant.

The sublimed compound is a crystalline canary-yellow substance; the crude product may vary in color from bright to creamy yellow, but since no differences in the analytical data and in i.r. spectrum between the sublimed substance and the crude product are observed, the sublimation is not recommended. Dodecacarbonyltetrairidium is only very slightly soluble in all solvents. A saturated methylene chloride solution shows an i.r. spectrum[6] which has bands at 2067 and 2027 cm.$^{-1}$ (1-cm. cell, Perkin Elmer 621). The i.r. spectrum of the solid (KBr pellets) has bands at 2084, 2053, and 2020 cm.$^{-1}$. The carbonyl is inert to air and water. It reacts with alkali hydroxides and cyanides, most rapidly in methanol or ethanol suspension, as well as with metallic sodium in tetrahydrofuran, giving

---

* In 24 hours about 3–4 mg. sublimed product could be obtained from 50 mg. of crude product.

a series of polynuclear carbonyliridates[7,8] such as $[Ir_6(CO_{15}]^{2-}$, $[Ir(CO)_4]^-$, and the known $[Ir_4(CO)_{11}H]^-$ and $[Ir_4(CO)_{10}]^{2-}$.

References

1. W. Hieber and H. Lagally, *Z. Anorg. Allgem. Chem.*, **245**, 321 (1940).
2. W. Hieber, H. Lagally, and A. Mayr, *ibid.*, **246**, 138 (1941).
3. S. H. H. Chaston and F. G. A. Stone, *Chem. Commun.*, **1967**, 964.
4. G. B. Kauffman and L. A. Teter, *Inorganic Syntheses*, **8**, 225 (1966).
5. M. Delepine, *Ann. Chim. (Paris)* (**9**)**7**, 277 (1917).
6. F. Cariati, V. Valenti, and G. Zerbi, *Inorg. Chim. Acta*, **3**, 378 (1969).
7. Work in progress in this laboratory.
8. L. Malatesta and G. Caglio, *Chem. Commun.*, **1967**, 420.

19. DISODIUM HEXAALKOXY-OCTA-μ₃-CHLORO-HEXAMOLYBDATES

Submitted by PIERO NANNELLI* and B. P. BLOCK*
Checked by D. LISSY† and B. B. WAYLAND†

Derivatives of molybdenum(II) chloride, in which the ligand chloride ions are replaced by alkoxy groups, can be readily prepared by treatment of molybdenum(II) chloride with excess sodium alkoxides in alcohol.[1]   The resulting crystalline products are sodium salts in which the anion consists of the octachloro-hexamolybdenum(II) group $[Mo_6Cl_8]^{4+}$ surrounded by six ligand alkoxide ions.   The bromide analog can be prepared starting from molybdenum(II) bromide and alcoholic sodium alkoxides by a similar procedure.[1] Presumably, molybdenum(II) iodide reacts in the same way.

In some cases alcohol interchange is a convenient preparative method.   As an example of this procedure, the preparation of

* Technological Center, Pennsalt Chemicals Corporation, King of Prussia, Pa. 19406.
† University of Pennsylvania, Philadelphia, Pa. 19104.

CENTRAL METHODIST COLLEGE LIBRARY
FAYETTE, MISSOURI   65248

disodium hexaalkoxy-octa-μ_3-chloro-hexamolybdates, from the corresponding methoxide and ethanol, is presented.

Because of their reactivity these derivatives are useful intermediates in the synthesis of a variety of compounds containing the halomolybdenum(II) cluster.   Individual alkoxy derivatives can be synthesized in less than $1\frac{1}{2}$ days.

A. DISODIUM OCTA-μ_3-CHLORO-HEXAMETHOXYHEXAMOLYBDATE(II)

$$(Mo_6Cl_8)Cl_2Cl_{4/2} + 6NaOCH_3 \rightarrow Na_2[(Mo_6Cl_8)(OCH_3)_6] + 4NaCl$$

Procedure

During all operations care should be taken to avoid atmospheric moisture.   A continuous, slow stream of dry nitrogen through the reaction flask is recommended.   The methanol used in the experiment should be anhydrous, preferably distilled over magnesium methoxide, and the ether dried with fresh sodium wire.

A 500-ml., three-necked flask is equipped with a magnetic stirring bar, a 250-ml. dropping funnel with side arm, and a reflux condenser with a Drierite-filled tube attached to it. Molybdenum(II) chloride[2] (60 g., 59.9 mmoles) is placed in the flask together with 50 ml. of methanol.   A solution of sodium methoxide in methanol is prepared directly in the dropping funnel by gradually adding a total of 11.5 g. (0.5 mole) of small pieces of sodium to 150 ml. of methanol in the funnel.   The solution is then added to the slurry of molybdenum(II) chloride and methanol over a 2–3-minute period.   The resulting yellow-orange suspension, while being stirred magnetically, is heated to reflux and then filtered hot through a fritted glass of medium porosity in order to remove the precipitated sodium chloride. Cooling the clear solution in an ice bath causes the methoxide derivative to separate as a yellow crystalline compound.   After about 2 hours the precipitate is collected on a fritted glass of coarse porosity and washed with two 50-ml. portions of ether–methanol solution (9:1 ratio), and then with ether. The product is dried at 50° under vacuum. Yield is *ca.* 38 g. (58%). *Anal.* Calcd. for $Na_2[(Mo_6Cl_8)(OCH_3)_6]$: C, 6.60; H, 1.66; Cl, 25.99; Mo, 52.74; Na, 4.21. Found: C, 6.56; H, 1.77; Cl, 25.8; Mo, 52.5; Na, 4.1.

B. DISODIUM OCTA-μ₃-CHLORO-HEXAETHOXYHEXAMOLYBDATE(II)

$$[Mo_6Cl_8]Cl_2Cl_{4/2} + 6NaOC_2H_5 \rightarrow Na_2[(Mo_6Cl_8)(OC_2H_5)_6] + 4NaCl$$

Procedure

Absolute ethanol* (175 ml.) is placed in a 500-ml., three-necked flask equipped with a gas inlet tube and a reflux condenser protected with a Drierite-filled drying tube. Small pieces of sodium (11.5 g., 0.5 mole) are then carefully added, and eventually the reaction mixture is heated in order to speed up dissolution of the sodium. A slow stream of dry nitrogen is passed through the flask during the entire reaction. After all the sodium has dissolved, the resulting solution of sodium ethoxide is cooled to room temperature. Then, over a 2–3-minute period, it is added with stirring to a slurry of 60 g. of molybdenum(II) chloride[2] (59.9 mmoles) in 50 ml. of absolute ethanol contained in a separate 500-ml., three-necked flask equipped with a mechanical stirrer and a reflux condenser to which is attached a Drierite-filled drying tube. In order to avoid exposure of the reagents to atmospheric moisture, the addition is conveniently made through an elbow connecting the two flasks while a stream of nitrogen is passed through the entire

---

* ■ *Note. If commercial absolute ethanol is used, it is recommended that the residual water be removed by treatment first with sodium and then with ethyl phthalate, with subsequent distillation.*

system.   Atmospheric moisture should also be avoided during all subsequent operations, preferably by working under nitrogen. The resulting yellow-orange suspension is stirred at 40–50° for one hour and then filtered through a fritted glass of medium porosity.   Upon addition of 600 ml. of anhydrous ether yellow crystals of the ethoxide derivative separate.   Separation of the product is favored by occasional scratching of the walls of the container.   After about 2 hours the precipitate is collected on a fritted glass of coarse porosity and washed with two 50-ml. portions of ether–ethanol solution (9:1 ratio) and then with ether. The compound is dried under vacuum at 50°.   Yield is 39.5 g. (55.9%).   *Anal.* Calcd. for $Na_2[(Mo_6Cl_8)(OC_2H_5)_6]$: C, 12.25; H, 2.57; Cl, 24.12; Mo, 48.96; Na, 3.91.   Found: C, 12.4; H, 2.47; Cl, 24.6; Mo, 48.2; Na, 4.0.

C. DISODIUM OCTA-μ_3-CHLORO-HEXAETHOXYHEXAMOLYBDATE(II)

(By Alcohol Interchange)

$$Na_2[(Mo_6Cl_8)(OCH_3)_6] + 6C_2H_5OH \rightarrow$$
$$Na_2[(Mo_6Cl_8)(OC_2H_5)_6] + 6CH_3OH$$

Procedure

The absolute ethanol used in this experiment should be freed from residual water as described in Sec. B.   Both toluene and ether are dried with fresh sodium wire.

A 500-ml., three-necked flask is equipped with a magnetic stirring bar, gas inlet tube, and a vacuum-jacketed fractionation column, 48 cm. long and 2.5 cm. wide, packed with glass helices, to which is attached a distilling head.   The flask is then charged with 10.8 g. of $Na_2[(Mo_6Cl_8)(OCH_3)_6]$* and 250 ml. of absolute ethanol.   The resulting solution, while being stirred magnetically, is heated to reflux by means of an oil bath held at 140–150°.   Nitrogen (about 30 bubbles per minute) is passed through the apparatus during this and subsequent operations.

* See Sec. A of this procedure.

After a one-hour period of reflux, distillation is started at the rate of about 20 ml. of distillate per hour, until a total of 170 ml. of liquid has been collected. To the residual solution is added 200 ml. of toluene, and the alcohol is distilled off completely as an azeotrope with toluene. Eventually a yellow suspension is obtained. The yellow crystalline solid is collected on a fritted glass of medium porosity and washed with ether. Care should be taken not to expose the compound to atmospheric moisture. The compound is dried under vacuum at 50°, giving a yield of 11.1 g. (95.4%). *Anal.* Calcd. for $Na_2[(Mo_6Cl_8)(OC_2H_5)_6]$: C, 12.25; H, 2.57. Found: C, 12.81; H, 2.81.

Properties

Both $Na_2[(Mo_6Cl_8)(OCH_3)_6]$ and $Na_2[(Mo_6Cl_8)(OC_2H_5)_6]$ are yellow, crystalline, diamagnetic solids, very soluble in alcohols and insoluble in inert solvents.[1] They are readily soluble in water with hydrolysis of the alkoxy groups. Complete hydrolysis by exposure to the atmospheric moisture, however, may require up to 1 week. Treatment with hydrochloric acid gives crystalline $(H_3O)_2[(Mo_6Cl_8)Cl_6]\cdot6H_2O$, from which the starting $[Mo_6Cl_8]Cl_2Cl_{4/2}$ can be obtained upon heating under vacuum.[3] Mixed molybdenum(II) halides,[3] $[Mo_6Cl_8]Br_2Br_{4/2}$ and $[Mo_6Cl_8]\text{-}I_2I_{4/2}$, can be prepared by heating under vacuum the corresponding $(H_3O)_2[(Mo_6Cl_8)Br_6]\cdot6H_2O$ and $(H_3O)_2[(Mo_6Cl_8)I_6]\cdot6H_2O$, obtained upon treatment of the alkoxides with hydrobromic or hydriodic acid, respectively. The compounds are decomposed by strong bases to hydroxides of molybdenum in higher oxidation states.

References

1. P. Nannelli and B. P. Block, *Inorg. Chem.*, **7**, 2423 (1968).
2. P. Nannelli and B. P. Block, *Inorganic Syntheses*, **12**, 170 (1970).
3. J. C. Sheldon, *J. Chem. Soc.*, **1960**, 1007.

Chapter Four

PHOSPHINE AND PHOSPHITE COMPLEXES OF LOW-VALENT METALS

20. TRIARYL PHOSPHITE COMPLEXES OF COBALT, NICKEL, PLATINUM, AND RHODIUM

Submitted by J. J. LEVISON* and S. D. ROBINSON*
Checked by J. G. VERKADE†

Organophosphine derivatives of the group VIII metals are frequently prepared by direct reaction of the phosphine ligand with a metal salt in refluxing alcoholic media.[1]   Attempts to prepare triaryl phosphite derivatives by similar routes have usually been frustrated by solvolysis of the ligands and formation of unisolable or uncharacterizable products.   These preparative difficulties have resulted in the relative neglect of this group of potentially valuable ligands.   However the data that are available in the literature suggest that the synthesis of a wide range of triaryl phosphite complexes containing metal hydride groups[2] or metals in low formal oxidation states[3] may be possible.   In this context it is interesting to note the close parallel, now becoming appar-

* University of London, King's College, Strand, London, W.C. 2, England.
† Iowa State University of Science and Technology, Ames, Iowa 50010.

ent, between the ligand properties of triphenyl phosphite, $P(OPh)_3$,[2,3] and phosphorus trifluoride, PF_3.[4]

The preparations described here are developed from published work by Malatesta et al.[5] and from more recent studies in the contributors' own laboratory.[2] The cobalt and nickel complexes are prepared by reduction of the corresponding metal nitrates with sodium tetrahydroborate in the presence of excess ligand, whereas the syntheses of the rhodium and platinum complexes involve simple ligand exchange processes. The preparative routes are suitable for use with triphenyl- or p-substituted triphenyl phosphites; reactions involving o- or m-substituted triphenyl phosphites give much reduced yields of products which are difficult to crystallize and are very air-sensitive. These features probably reflect the unfavorable stereochemistry of the o- and m-substituted ligands.

The cobalt complexes described here, together with the triethyl phosphite analog,[6-8] are the only examples of simple cobalt phosphite hydride complexes reported to date and were the first examples of metal hydrides stabilized by phosphite ligands.

Tetrakis(triaryl phosphite)nickel(0) complexes are well known and are among the most stable nickel(0) derivatives to be reported.[3] They have previously been synthesized by reaction of triaryl phosphites with bis(cyclopentadienyl)nickel[9] and nickel tetracarbonyl.[10] Syntheses involving reduction of nickel(II) salts with various reducing agents including organoaluminum compounds[11] and sodium dithionite[12] have also been used. Tetrakis(trialkyl phosphite)nickel(0) complexes have been prepared by several of these routes and by reaction of nickel(II) salts with potassium graphite[13] or aqueous triethylamine[14] in the presence of excess trialkyl phosphite. Tris- and bis(triaryl phosphite)nickel(0) complexes have been synthesized by treatment of bis(acrylonitrile)nickel(0) with stoichiometric amounts of the triaryl phosphite.[15] The route described below is the most convenient available for the preparation of tetrakis-

(triaryl phosphite)nickel(0) complexes but is not suitable for the preparation of their trialkyl phosphite analogs.

Tris- and tetrakis(triaryl phosphite)platinum(0) complexes have been prepared by reduction of platinum(II) phosphite derivatives with hydrazine.[5] The tetrakis complexes have also been prepared by ligand exchange processes, and the synthesis described here is based on this latter procedure. The chemistry of platinum phosphite complexes has not been extensively studied.

The hydridotetrakis(triphenyl phosphite)rhodium complex described below is the first example of a rhodium hydride complex stabilized by phosphite ligands.[2]

■ *Note. The triaryl phosphite complexes, with the possible exception of the platinum derivative, appear to interact with oxygen when prepared or manipulated in the presence of air. This interaction, which leads to large fluctuations in melting behavior,* does not cause any other apparent changes. However, in view of these observations, it is recommended that all preparations and subsequent manipulations of the complexes be performed in a nitrogen-filled dry-box.*

Small fluctuations in melting behavior appear to be an intrinsic property of the complexes even when they are prepared and purified in the absence of air. Thus, melting-point data are of no value as a criterion of identity or purity for these complexes and are therefore not recorded below.

A. HYDRIDOTETRAKIS(TRIPHENYL PHOSPHITE)COBALT(I)

$$Co(NO_3)_2 \cdot 6H_2O + 4P(OPh)_3 \xrightarrow[\text{ethanol}]{NaBH_4} [CoH\{P(OPh)_3\}_4]$$

■ *Note. See the recommendation in the note above concerning the use of inert atmosphere.*

---

* The contributors thank the checker for valuable observations concerning the melting behavior of these complexes.

Procedure

Triphenyl phosphite (15.5 g., 0.050 mole) is added to a solution of cobalt(II) nitrate hexahydrate (2.91 g., 0.010 mole) in ethanol (60 ml.). Sodium tetrahydroborate (1.0 g.) is dissolved in warm ethanol (25 ml.) and the solution cooled rapidly to room temperature. The sodium tetrahydroborate solution is added dropwise over a period of 10 minutes to the well-stirred cobalt nitrate solution. The purple color of the cobalt(II) salt rapidly discharges, and a pale yellow precipitate deposits. After stirring for a further 10 minutes the product is filtered off, washed with ethanol, and dried *in vacuo*. Yield is 11.0 g. (85%). The product may be further purified by dissolving in the minimum volume of benzene and reprecipitating by dropwise addition of methanol. *Anal.* Calcd. for $C_{72}H_{61}CoO_{12}P_4$: C, 66.48; H, 4.74; P, 9.52. Found: C, 66.60; H, 4.47; P, 9.21%. The tri-*p*-tolyl phosphite complex can be prepared (85% yield) by an analogous route. *Anal.* Calcd. for $C_{84}H_{85}CoO_{12}P_4$: C, 68.64; H, 5.84. Found: C, 68.74; H, 5.92%.

B. TETRAKIS(TRIPHENYL PHOSPHITE)NICKEL(0)

$$Ni(NO_3)_2 \cdot 6H_2O + 4P(OPh)_3 \xrightarrow[\text{ethanol}]{NaBH_4} [Ni\{P(OPh)_3\}_4]$$

■ *Note.* *See the recommendation in the note on page 107 concerning the use of inert atmosphere.*

*Procedure**

The procedure is identical with that of Sec. A except for the replacement of nickel(II) nitrate hexahydrate for the cobalt salt. [The formula weights of $Co(NO_3)_2 \cdot 6H_2O$ and $Ni(NO_3)_2 \cdot 6H_2O$ are virtually identical.] Yield is 12.0 g. (92%). Further purification of the complex may be readily effected by reprecipitation from benzene solution using methanol. *Anal.*

* Compare synthesis 21D, page 116.

Calcd. for $C_{72}H_{60}NiO_{12}P_4$: C, 66.53; H, 4.65; P, 9.54. Found: C, 66.80; H, 4.62; P, 9.27%. The tri-p-tolyl phosphite can be prepared (80% yield) by an analogous route. *Anal.* Calcd. for $C_{84}H_{84}NiO_{12}P_4$: C, 68.70; H, 5.77. Found: C, 69.06; H, 5.83%.

C. TETRAKIS(TRIPHENYL PHOSPHITE)PLATINUM(0)

$$Pt(PPh_3)_4 + 4P(OPh)_3 \xrightarrow{\text{benzene}} [Pt\{P(OPh)_3\}_4] + 4PPh_3$$

■ *Note.* See the recommendation in the note on page 107 concerning the use of inert atmosphere.

Procedure

A solution of tetrakis(triphenylphosphine)platinum (1.24 g., 0.001 mole) [*Inorganic Syntheses*, **11**, 105 (1968)] in benzene (10 ml.) is added to a solution of triphenyl phosphite (1.55 g., 0.005 mole) in benzene (5 ml.). The resultant pale-colored solution is filtered, diluted with n-hexane (25 ml.), and cooled overnight at 5°. The white crystalline product is filtered off, extracted with warm n-hexane (25 ml.) to remove coprecipitated triphenylphosphine, and dried *in vacuo.* Yield is 1.2 g. (84%). The product may be further purified by dissolving in the minimum volume of benzene and reprecipitating by addition of n-hexane (m.p. 148–154°). *Anal.* Calcd. for $C_{72}H_{60}O_{12}P_4Pt$: C, 60.20; H, 4.22; P, 8.63. Found: C, 60.45; H, 4.37; P, 8.35%.

D. HYDRIDOTETRAKIS(TRIPHENYL PHOSPHITE)RHODIUM(I)

$$RhH(CO)(PPh_3)_3 + 4P(OPh)_3 \xrightarrow{\text{ethanol}} [RhH\{P(OPh)_3\}_4] + 3PPh_3 + CO$$

■ *Note.* See the recommendation in the note on page 107 concerning the use of inert atmosphere.

Procedure

Triphenyl phosphite (1.55 g., 0.005 mole) is added to a suspension of carbonylhydridotris(triphenylphosphine)rhodium[16] (0.92 g., 0.001 mole) in ethanol (30 ml.), and the mixture is heated gently until the complex is dissolved and a pale yellow solution formed. The hot solution is filtered to remove any particles of unreacted complex, and the filtrate is cooled slowly to room temperature and thence at 5° overnight. (Rapid cooling or use of higher concentrations of complex results in deposition of an intractable oil.) The resultant pale yellow solid is filtered off, washed with ethanol and then *n*-hexane (to remove triphenylphosphine), and dried *in vacuo*. The complex is purified by dissolving in the minimum volume of cold methylene chloride and precipitating by *dropwise* addition of methanol. Formation of crystalline product is greatly aided by careful scratching of the sides of the recrystallization vessel with a spatula. Yield is 0.60 g. (45%). *Anal.* Calcd. for $C_{72}H_{61}O_{12}P_4Rh$: C, 64.30; H, 4.50; P, 9.23. Found: C, 63.63; H, 4.48; P, 9.07%.

Properties

The cobalt and rhodium complexes are pale yellow in color, whereas those of nickel and platinum are white. The rhodium derivative decomposes in the presence of air, but the other complexes apparently remain intact for prolonged periods. However, as noted on page 107, significant changes in the melting behavior of the cobalt, nickel, and rhodium complexes following brief exposure to air suggest that some form of interaction with oxygen or moisture is in fact occurring. Osmometric molecular-weight data indicate that the cobalt and nickel complexes are undissociated in benzene solution, whereas those of rhodium and platinum apparently undergo extensive ligand dissociation under similar conditions. (Molecular weights: $[Pt\{P(OPh)_3\}_4]$

requires 1435, found: 561; [RhH{P(OPh)$_3$}$_4$] requires 1345, found: 677.) The complexes, with the possible exception of the rhodium derivative, are much less reactive than their triphenylphosphine analogs, where these are known, and do not readily undergo substitution reactions. The cobalt derivatives, by comparison with other known cobalt hydride complexes, are remarkably stable.

X-ray data suggest that the complexes [CoH{P(OPh)$_3$}$_4$], [Ni{P(OPh)$_3$}$_4$], [Pt{P(OPh)$_3$}$_4$], and [RhH{P(OPh)$_3$}$_4$] are isomorphous. A tetrahedral arrangement of phosphite ligands about the metal atom is to be expected for the nickel(0) and platinum(0) complexes and on the evidence of the x-ray data appears highly probable for the cobalt and rhodium hydride derivatives also. The hydride ligand in the latter complexes is probably located along a C_3 axis of the molecule as postulated for the related complex [CoH(CO)$_4$].

No ν_{M-H} stretching vibrations can be detected in the infrared spectra of the cobalt and rhodium complexes, but the presence of hydride ligands is confirmed by the appearance of a quintet pattern in the high-field n.m.r. spectra of these derivatives. The apparent equivalent coupling of the hydride ligand to the four ^{31}P nuclei has been tentatively explained in terms of a nonrigid structure.[2]

References

1. G. Booth, *Advan. Inorg. Chem. Radiochem.*, **6**, 1 (1964).
2. J. J. Levison and S. D. Robinson, *Chem. Commun.*, **1968**, 1045.
3. L. Malatesta, R. Ugo, and S. Cenini, Advances in Chemistry Series, no. 62, 318, 1967.
4. T. Kruck, *Angew. Chem. (Intern. Ed.)*, **6**, 53 (1967).
5. L. Malatesta and C. Cariello, *J. Chem. Soc.*, **1958**, 2323.
6. R. D. Mullineaux, U.S. Patent 3,290,318; *C. A.*, **66**, 65080 (1967).
7. W. Kruse and R. H. Atalla, *Chem. Commun.*, **1968**, 921.
8. M. E. Volpin and I. S. Kolomnikov, Katal. Reakts. Zhidk, Faze Tr. Vses. Konf., 2d, Alma-Ata, Kaz. S.S.R., 429, 1966; C. A., **69**, 46340 (1968).
9. J. R. Olechowski, C. G. McAlister, and R. F. Clark, *Inorganic Syntheses*, **9**, 181 (1967).
10. L. S. Meriwether and J. R. Leto, *J. Am. Chem. Soc.*, **83**, 3192 (1961).

11. R. F. Clark, C. D. Storrs, and C. G. McAlister, Belgian Patent, 621,207; *C. A.*, **59**, 11342 (1963).
12. L. G. Cannell, U.S. Patent 3,102,899; *C. A.*, **60**, 1062 (1964).
13. K. A. Jensen, B. Nygaard, G. Elisson, and P. H. Nielson, *Acta. Chem. Scand.*, **19**, 768 (1965).
14. R. S. Vinal and L. T. Reynolds, *Inorg. Chem.*, **3**, 1662 (1964).
15. N. V. Kutepow, H. Seibt, and F. Meier, German Patent 1,144,268; *C. A.*, **59**, 3790 (1963).
16. D. Evans, G. Yagupsky, and G. Wilkinson, *J. Chem. Soc.(A)*, **1968**, 2660.

21. TETRAKIS(TRIETHYL PHOSPHITE)NICKEL(0), PALLADIUM(0), AND PLATINUM(0) COMPLEXES

Submitted by MAX MEIER* and FRED BASOLO*
Checked by W. R. KROLL,† D. MOY,† and M. G. ROMANELLI†

A. TETRAKIS(TRIETHYL PHOSPHITE)NICKEL(0)

$$NiCl_2 \cdot 6H_2O + 5P(OC_2H_5)_3 + 2(C_2H_5)_2NH \rightarrow$$
$$[Ni\{P(OC_2H_5)_3\}_4] + (C_2H_5O)_3PO + 2(C_2H_5)_2NH_2Cl + 5H_2O$$

Procedure

The triethyl phosphite (Eastman) used in these preparations was distilled in a vacuum prior to use, b.p. 51° at 13 mm. Hg. The preparations were carried out in air; an inert atmosphere was not necessary.

The nickel(0) compound can be prepared by the method of Vinal and Reynolds.[1]   Nickel(II) chloride hexahydrate (5.0 g., 0.021 mole) is dissolved in 100 ml. of methanol contained in a 250-ml. round-bottomed flask.   A stirring bar is placed in the flask, and it is put in an ice–water bath standing on a magnetic stirrer.   The solution is allowed to cool with stirring for 10 minutes, and 18 ml. of triethyl phosphite are then added over a

* Department of Chemistry, Northwestern University, Evanston, Ill. 60201.
† Esso Research and Engineering Company, Corporate Research Laboratories, Linden, N.J. 07036.

period of 1 minute. Upon addition of $P(OC_2H_5)_3$ the solution turns dark red. With further cooling and vigorous stirring diethylamine is added dropwise from a syringe containing 5 ml. of $(C_2H_5)_2NH$ over a period of 10 minutes. After about half the diethylamine has been added, white crystals of $[Ni\{P-(OC_2H_5)_3\}_4]$ appear. Addition of diethylamine is stopped when the color of the liquid has faded to pink. [Further addition of diethylamine, until the liquid phase is yellow or green will cause contamination of the product with nickel(II) compounds.] The crystals are collected on a glass frit by means of a suction filter and washed with methanol which has previously been cooled in an ice–water bath. The washing is continued until the product is colorless. The product is transferred quickly to a drying vessel (e.g., a 50-ml. round-bottomed flask) which is connected to a liquid-nitrogen-cooled trap and a vacuum pump (our pump was capable of producing a vacuum of 10^{-2} mm. Hg) and dried for 5 hours at room temperature. The yield is typically about 40% using methanol as solvent. The checkers report that using acetonitrile yields of 55–60% can be obtained. *Anal.* Calcd. for $C_{24}H_{60}O_{12}P_4Ni$: C, 39.83; H, 8.30. Found: C, 40.22; H, 8.58.

Properties

Tetrakis(triethyl phosphite)nickel(0) can be handled in air. It is best stored in an evacuated and sealed tube. On exposure to air for several hours the substance turns green. The compound is insoluble in water, somewhat soluble in methanol, and very soluble in hydrocarbons. It does not dissociate in hydrocarbon solutions.

B. TETRAKIS(TRIETHYL PHOSPHITE)PALLADIUM(0)

$$K_2PdCl_4 + 5P(OC_2H_5)_3 + 2(C_2H_5)_2NH + H_2O \rightarrow$$
$$[Pd\{P(OC_2H_5)_3\}_4] + (C_2H_5O)_3PO + 2(C_2H_5)_2NH_2Cl + 2KCl$$

Procedure

A concentrated solution of potassium tetrachloropalladate(II), K_2PdCl_4 (0.330 g., 0.001 mole), is prepared by dissolving it in a minimum amount (*ca.* 2.5 ml.) of water at room temperature. A methanol (3 ml.) solution of triethyl phosphite (0.831 g., 0.005 mole) is placed in a test tube containing a small stirring bar. The triethyl phosphite (Matheson, Coleman and Bell) was distilled before use and kept under nitrogen (b.p. 77°/25 mm). ■ *Note. It is very important that the amount of triethyl phosphite not exceed the stoichiometrically required 0.005 mole.* The test tube is placed in an ice–water bath standing on a magnetic stirrer, and the aqueous solution of chloropalladate is added with stirring.* A yellow solution results. If two liquid phases are formed, methanol is added dropwise until the solution is homogeneous. The yellow solution is cooled with stirring for 2 minutes, and then 0.21 ml. of diethylamine is added to the solution from a small syringe, with further cooling and vigorous stirring. (The tip of the needle is immersed in the solution.) A white precipitate forms immediately, which is collected on a glass frit and washed quickly with a few milliliters of water. The product is rapidly transferred to a drying flask which is connected to a liquid-nitrogen-cooled trap and a vacuum pump and dried for one hour at room temperature, then for an additional 3 hours at 0°. The compound melts with decomposition at 112° under nitrogen, but at much lower temperatures in air. *Anal.* Calcd. for $C_{24}H_{60}O_{12}P_4Pd$: C, 37.35; H, 7.87. Found: C, 36.48; H, 7.84.

Properties

The dry product, $[Pd\{P(OC_2H_5)_3\}_4]$, containing no adsorbed triethyl phosphite is air-sensitive, turning black on exposures

*■ *Note.* The checkers recommend that the entire preparative procedure be carried out in an apparatus flushed with nitrogen.

to air exceeding a few minutes. It should be handled in an inert atmosphere (glove box or a large beaker filled with argon). It is stored in an evacuated and sealed tube. The complex is insoluble in water, soluble in methanol, and very soluble in hydrocarbons, in which it does not dissociate.

C. TETRAKIS(TRIETHYL PHOSPHITE)PLATINUM(0)

$$K_2PtCl_4 + 5P(OC_2H_5)_3 + 2KOH \rightarrow$$
$$[Pt\{P(OC_2H_5)_3\}_4] + (C_2H_5P)_3PO + 4KCl + H_2O$$

Procedure

The platinum(0) compound can be prepared by a method analogous to that of Malatesta and Cariello.[2] Powdered potassium hydroxide* (0.350–0.400 g., 0.006 mole) is dissolved in 10 ml. of methanol contained in a large test tube (about 1 in. diameter). To this is added triethyl phosphite (2.5 g., 0.015 mole) and a small stirring bar, and then the test tube is placed in an oil bath kept at 75° by a heater-stirrer plate. When the solution in the test tube has reached the temperature of 60°, a solution of potassium tetrachloroplatinate(II), K_2PtCl_4 (1.24 g., 0.003 mole), in about 20 ml. of water is slowly added with stirring. Immediately or within a few minutes, colorless crystals separate. The crystals are collected on a glass frit, washed with a few milliliters of an ethanol–water mixture (50% by volume) and dried under vacuum for 4 hours at room temperature. Yields vary from 0.45 to 0.58 g. (52–67%), m.p. 114°. *Anal.* Calcd. for $C_{24}H_{60}O_{12}P_4Pt$: C, 33.53; H, 7.03. Found: C, 31.0; H, 7.00.

Properties

The complex $[Pt\{P(OC_2H_5)_3\}_4]$ can be handled in air. It is stored in an evacuated and sealed tube. On exposure to air

* The weight of potassium hydroxide depends on the assay of the material available: 0.006 mole of 100% KOH weighs 0.337 g. The weight range given is representative.

exceeding several hours the substance turns black. The compound is insoluble in water, somewhat soluble in methanol, and very soluble in hydrocarbons in which it does not dissociate.

D. TETRAKIS(TRIPHENYL PHOSPHITE)NICKEL(0)

$$(C_5H_5)_2Ni + P(OC_6H_5)_3 \rightarrow [Ni\{P(OC_6H_5)_3\}_4]$$

Procedure*

This nickel(0) compound can be prepared by the method of Olechowski, McAlister, and Clark.[3] Dicyclopentadieneylnickel, $(C_5H_5)_2Ni$ (5.0 g., 0.027 mole), and triphenyl phosphite (50 g., 0.16 mole) are placed in a three-necked flask equipped with a thermometer, a nitrogen inlet, and a dropping funnel on which the nitrogen outlet is attached. The flask is heated in an oil bath to 90°, and a green solution is formed initially. If not all of the nickelocene dissolves, more triphenyl phosphite is added from the funnel. The solution turns brown in the course of the reaction. After 2 hours of heating the solution is allowed to cool to 50°, and the nitrogen inlet is disconnected. The colorless product is precipitated by adding acetone. The product is collected on a glass frit, washed with acetone, and dried in a vacuum. Yields are typically 96%. *Anal.* Calcd. C, 66.53; H, 4.65. Found: C, 66.65; H, 4.70.

The checkers found that use of a commercially available solution of nickelocene in toluene (*ca.* 0.27 M) is less desirable, but is admissible with certain modifications of the above procedure. If such a solution is used directly, the amount of $(C_6H_5O)_3P$ should be increased to 225 mmoles and a reaction time of 3 days allowed, to give a yield of 53%. Alternatively, the toluene may be vacuum-stripped prior to addition of the triphenyl phosphite. Again, an excess of phosphite (325 mmoles) and a longer reaction time (24 hours) were required, giving a yield of 69%.

* Compare synthesis 20B, page 108.

Properties

The complex [Ni{P(OC$_6$H$_5$)$_3$}$_4$] can be handled in air. The compound is insoluble in polar solvents and only slightly soluble in hydrocarbons.

References

1. R. S. Vinal and I. T. Reynolds, *Inorg. Chem.*, **3**, 1062 (1964).
2. L. Malatesta and C. Cariello, *J. Chem. Soc.*, **1958**, 2323.
3. J. R. Olechowski, C. G. McAlister, and R. F. Clark, *Inorg. Chem.*, **4**, 246 (1965); *Inorganic Syntheses*, **9**, 181 (1967).

22. LOW-VALENT METAL COMPLEXES OF DIETHYL PHENYLPHOSPHONITE

Submitted by D. TITUS,* A. A. ORIO,* and HARRY B. GRAY*
Checked by G. W. PARSHALL† and J. J. MROWCA†

The diethyl ester of phenylphosphonous acid (diethoxyphenyl-phosphine) provides an easy pathway to relatively stable tetrakis complexes of zero- and low-valent transition metals.[1,2] Anhydrous metal halides serve as the metal source for the complexes, avoiding the necessity of inconvenient starting materials such as nickel carbonyl. The nickel(0) complex is formed by reaction with the phosphonite in ethanol; with the addition of sodium tetrahydroborate, relatively stable dihydridoiron(II) and hydridocobalt(I) complexes are obtained.

The phosphonite is prepared from inexpensive starting materials, following the method of Rabinowitz and Pellon:[3]

$$\text{PhPCl}_2 + 2\text{C}_2\text{H}_5\text{OH} + 2\text{N(C}_2\text{H}_5)_3 \xrightarrow{\text{benzene}} \text{PhP(OC}_2\text{H}_5)_2 + 2(\text{C}_2\text{H}_5)_3\text{NHCl}$$

* Contribution no. 3869 from the Arthur Amos Noyes Laboratory of Chemical Physics, Californ'a Institute of Technology, Pasadena, Calif. 91109.
† Central Research Dept., Experimental Station, E. I. du Pont de Nemours & Company, Wilmington, Del. 19898.

This reaction is conveniently carried out in a 2-l., three-necked flask with 143 g. phenylphosphonous dichloride (0.8 mole), 74 g. absolute ethanol (1.6 moles), 163 g. triethylamine (1.6 moles), and 700 ml. benzene. The yield of water-white product is about 100 g. (63%). It is soluble in most organic solvents and slowly reacts with air; it should be stored under nitrogen. Its infrared spectrum has been reported.[4]

A. TETRAKIS(DIETHYL PHENYLPHOSPHONITE)NICKEL(0)

$$NiCl_2 + 5PhP(OC_2H_5)_2 \rightarrow [Ni\{PhP(OC_2H_5)_2\}_4] + Cl_2PhP(OC_2H_5)_2$$

Procedure

Anhydrous nickel(II) chloride (1.30 g., 0.01 mole) is dissolved in 100 ml. of absolute ethanol in a 200-ml. round-bottomed flask fitted with a reflux condenser and nitrogen inlet and outlet. Diethyl phenylphosphonite (10 g., 0.05 mole) is added, and the solution is heated to reflux. After 3 hours, the heat is removed and the solution is allowed to cool slowly to room temperature. The product separates as yellow crystals from the solution. With a stream of nitrogen passing through the flask, the mother liquor is transferred by syringe to another 200-ml. flask; the crystals are washed with two 20-ml. portions of absolute ethanol, and dried *in vacuo*. Concentration of the mother liquor to 30 ml. yields additional product. Yield is 8.2 g. (97%). *Anal.* Calcd. for $C_{40}H_{60}O_8NiP_4$: C, 56.51; H, 7.11; P, 14.60. Found: C, 56.27; H, 7.17; P, 14.54.

B. TETRAKIS(DIETHYL PHENYLPHOSPHONITE)- HYDRIDOCOBALT(I)

$$Co^{2+} + 4PhP(OC_2H_5)_2 \xrightarrow[\text{ethanol}]{\text{sodium tetrahydroborate}} [HCo\{PhP(OC_2H_5)_2\}_4]$$

Procedure

A 100-ml., round-bottomed, three-necked flask, equipped with reflux condenser, dropping funnel, magnetic stirring bar, and nitrogen inlet and outlet, is charged with 1.30 g. anhydrous cobalt chloride (0.01 mole) and 60 ml. absolute ethanol. The mixture is heated to 60–70° with stirring, and to the resultant dark-blue solution is added 10 g. diethyl phenylphosphonite (0.05 mole), yielding a dark green solution. A solution of sodium tetrahydroborate in ethanol (about 0.8 g. in 25 ml.) is added slowly from the dropping funnel until the color becomes bright yellow. The mixture is stirred for 15 minutes and then filtered under nitrogen using a medium porosity filter tube.[5] There should only be a small amount of fine precipitate in the reaction flask at this time; an increase in temperature and/or additional ethanol may be required to keep the product in solution for filtration. The filtrate is allowed to cool to room temperature; the product separates from solution as yellow-orange plates. These are washed with ethanol and dried *in vacuo.* Yield is 6.8 g. (80%). *Anal.* Calcd. for $C_{40}CoH_{61}O_8P_4$: C, 56.37; H, 7.16; P, 14.53. Found: C, 56.25; H, 7.18; P, 14.47.

C. TETRAKIS(DIETHYL PHENYLPHOSPHONITE)-DIHYDROIRON(II)

$$Fe^{2+} + 4PhP(OC_2H_5)_2 \xrightarrow{\text{sodium tetrahydroborate}} [FeH_2\{PhP(OC_2H_5)_2\}_4]$$

Procedure

In an apparatus identical to that used for the preparation of the cobalt complex are combined 1.27 g. anhydrous iron(II) chloride (0.01 mole), 10 g. diethyl phenylphosphonite (0.05 mole), and 40 ml. absolute ethanol. The mixture is heated to reflux under a slow stream of nitrogen. After 3 hours the heat is removed and a solution of sodium tetrahydroborate in absolute ethanol (0.2 g. in 10 ml.) is added dropwise with stirring.   A reddish color appears initially upon addition, but it fades rapidly and the color is dark yellow-brown when addition is complete. The mixture is filtered under nitrogen as with the cobalt complex, although extra heating is not necessary.   The filtered solution is allowed to stand at room temperature for 24 hours.   The product separates from solution as pale yellow prisms; it is sometimes necessary to store the solution in a refrigerator for a day or two to effect crystallization.   Yield is 4.6 g. (54%). *Anal.* Calcd. for $C_{40}FeH_{62}O_8P_4$: C, 56.48; H, 7.35; Fe, 6.57; P, 14.56.   Found: C, 56.46; H, 7.30; Fe, 6.72; P, 14.73.

Properties

All of the complexes described are diamagnetic.   The Ni(0) and Co(I) complexes are quite soluble in nonpolar organic solvents such as cyclohexane, and the Fe(II) complex is soluble in both polar and nonpolar organic solvents.   The Fe(II) complex is unchanged under nitrogen after several weeks, although it begins to decompose after only a few hours of exposure to air. The Ni(0) and Co(I) complexes are much more stable; decomposition occurs only after exposure to air for several days.

Absorption maxima of electronic spectral bands for the Ni(0) complex in cyclohexane solution are placed at 30,000 and 40,200 cm.$^{-1}$.   Characteristic infrared absorptions (not observed in the free phosphine ligand) for the hydrido complexes are found at 2017(w), 1950(w), 497(m), and 460(m) for cobalt(I) and 1975(w), 508(m), 500(m), 489(w), and 476(m) for iron(II) (Nujol mulls).

References

1. B. Chastain, E. Rick, R. Pruett, and H. B. Gray, *J. Am. Chem. Soc.*, **90**, 3994 (1968).
2. A. Orio, B. Chastain, and H. B. Gray, *Inorg. Chim. Acta*, **3**, 8 (1969).
3. R. Rabinowitz and J. Pellon, *J. Org. Chem.*, **26**, 4623 (1961).

4. L. Daasch and D. Smith, *Anal. Chem.*, **23**, 855 (1951).
5. A detailed description of filtration under nitrogen has been given by W. L. Jolly, *Inorganic Syntheses*, **11**, 117 (1968).

23. TETRAKIS(TRIPHENYLPHOSPHINE)PALLADIUM(0)

$$2PdCl_2 + 8P(C_6H_5)_3 + 5NH_2NH_2 \cdot H_2O \rightarrow$$
$$2[Pd\{P(C_6H_5)_3\}_4] + 4NH_2NH_2 \cdot HCl + N_2 + 5H_2O$$

Submitted by D. R. COULSON*
Checked by L. C. SATEK† and S. O. GRIM†

Preparation of phosphine and phosphite complexes of palladium(0) have been reported to result from reduction of palladium(II) complexes in the presence of the desired ligand.[1-5] The products are generally formulated as PdL_{4-n} (where $n = 0, 1$), depending upon the nature and amount of the ligand used. A related complex, $[Pd\{P(C_6H_5)_3\}_2]_n$, has also been reported.[6]

Although this preparation is similar in concept to these previous ones, advantage is gained in being able to obtain a high yield of $[Pd\{P(C_6H_5)_3\}_4]$ in one step from palladium dichloride.

Procedure

A mixture of palladium dichloride (17.72 g., 0.10 mole), triphenylphosphine (131 g., 0.50 mole), and 1200 ml. of dimethyl sulfoxide is placed in a single-necked, 2-l., round-bottomed flask equipped with a magnetic stirring bar and a dual-outlet adapter (Note 1). A rubber septum and a vacuum-nitrogen

* Central Research Department, Experimental Station, E. I. du Pont de Nemours & Company, Wilmington, Del. 19898.
† University of Maryland, College Park, Md. 20704.

system are connected to the outlets. The system is then placed under nitrogen with provision made for pressure relief through a mercury bubbler. The yellow mixture is heated by means of an oil bath with stirring until complete solution occurs (*ca.* 140°). The bath is then taken away, and the solution is rapidly stirred for approximately 15 minutes. Hydrazine hydrate (20 g., 0.40 mole) is then rapidly added over approximately 1 minute from a hypodermic syringe. A vigorous reaction takes place with evolution of nitrogen. The dark solution is then immediately cooled with a water bath; crystallization begins to occur at *ca.* 125°. At this point the mixture is allowed to cool without external cooling. After the mixture has reached room temperature it is filtered under nitrogen on a coarse, sintered-glass funnel. The solid is washed successively with two 50-ml. portions of ethanol and two 50-ml. portions of ether. The product is dried by passing a slow stream of nitrogen through the funnel overnight. The resulting yellow crystalline product weighs 103.5–108.5 g. (90–94% yield).

A melting point determination (Note 2) on a sample in a sealed capillary tube under nitrogen gave a decomposition point of 116° (uncorrected). This compares with a similar determination (115°) performed on the product prepared by the method of Malatesta and Angoletta.[1] *Anal.* Calcd. for $C_{72}H_{60}PdP_4$: C, 75.88; H, 5.25; P, 10.75.   Found: C, 75.3; H, 5.36; P, 10.7.

Properties

The $[Pd\{P(C_6H_5)_3\}_4]$ complex obtained by this procedure is a yellow, crystalline material possessing moderate solubilities in benzene (50 g./l.), methylene chloride, and chloroform. The compound is less soluble in acetone, tetrahydrofuran and acetonitrile. Saturated hydrocarbon solvents give no evidence of solution. Although the complex may be handled in air, it is best stored under nitrogen to ensure its purity.

In benzene, molecular-weight measurements suggest substantial dissociation:[1,4]

$$[Pd\{P(C_6H_5)_3\}_4] \rightleftarrows [Pd\{P(C_6H_5)_3\}_{4-n}] + nP(C_6H_5)_3$$

Solutions of the complex in benzene rapidly absorb molecular oxygen giving an insoluble, green oxygen complex, $[Pd\{P(C_6H_5)_3\}_2O_2]$.[7] This oxygen complex has been implicated as an intermediate in a catalytic oxidation of phosphines.[2,8]

Related displacements with acetylenes[9] and electrophilic olefins[6] have been reported to give complexes formulated as $[Pd\{P(C_6H_5)_3\}_2$ (olefin or acetylene)]. Also, oxidative additions of alkyl and aryl halides have been shown to occur giving palladium(II) complexes, $[Pd\{P(C_6H_5)_3\}_2(R)Cl]$.[10]

As a catalyst, the complex has been shown capable of dimerizing butadiene to give 1,3,7-octatriene.[11]

Notes

1. The checkers worked on one-third of the stated scale, obtaining a yield of 37.4 g. (97%).

2. The checkers report that decomposition temperature does not appear to be a good criterion of identity or purity since it is not very reproducible.

References

1. L. Malatesta and M. Angoletta, *J. Chem. Soc.*, **1957**, 1186.
2. S. Takahashi, K. Sonogashira, and N. Hagihara, *Nippon Kagaku Zasshi*, **87**, 610 (1966); *C. A.*, **65**, 14485 (1966).
3. T. Kruck and K. Baur, *Angew. Chem.*, **77**, 505 (1965).
4. E. O. Fischer and H. Werner, *Chem. Ber.*, **95**, 703 (1962).
5. J. Chatt, F. A. Hart, and H. R. Watson, *J. Chem. Soc.*, **1962**, 2537.
6. P. Fitton and J. E. McKeon, *Chem. Commun.*, **1968**, 4.
7. C. J. Nyman, C. T. Wymore, and G. Wilkinson, *J. Chem. Soc.* (A), **1968**, 561.
8. G. Wilke, H. Schott, and P. Heimbach, *Angew. Chem.* (*Intern. Ed.*), **6**, 92 (1967).
9. S. Takahashi and N. Hagihara, *J. Chem. Soc. Japan* (Pure Chem. Sec.), **88**, 1306 (1967).

10. P. Fitton, M. P. Johnson, and J. E. McKeon, *Chem. Commun.*, **1968**, 6.
11. S. Takahashi, T. Shibano, and N. Hagihara, *Bull. Chem. Soc. Japan*, **41**, 454 (1968).

24. TETRAKIS(TRIPHENYLPHOSPHINE)NICKEL(0)

$$Ni(C_5H_7O_2)_2 + 2(C_2H_5)_3Al + 4(C_6H_5)_3P \rightarrow$$
$$[\{(C_6H_5)_3P\}_4Ni] + 2(C_5H_7O_2)Al(C_2H_5)_2 + 2C_2H_4 + \cdots$$

Submitted by R. A. SCHUNN*
Checked by E. C. ASHBY† and J. DILTS†

This procedure is based on that previously described by Wilke.[1] The use of $NaBH_4$ and sodium naphthalide as reducing agents has been more recently reported by Ugo.[2]

Proecdure

The entire procedure, including purification, is performed in an anhydrous, oxygen-free atmosphere using anhydrous, deoxygenated solvents (Note 1). A 2-l., four-necked, round-bottomed flask is equipped with a mechanical stirrer, 250-ml. pressure-equalizing dropping funnel, thermometer, and T tube, one side of which is attached to a source of dry nitrogen and the other side to a silicone oil bubbler. The flask is charged with 21.3 g. (0.083 mole) of anhydrous bis(2,4-pentanedionato)nickel (Note 2) and 125 g. (0.48 mole) of triphenylphosphine. The flask is thoroughly flushed with nitrogen (by removing the stopper from the addition funnel), and 800 ml. of diethyl ether is then added through the funnel. The green slurry is stirred and cooled to 0° with an ice–methanol bath. A solution of 28.0 g. (0.245 mole)

* Central Research Department, Experimental Station, E. I. du Pont de Nemours & Company, Wilmington, Del. 19898.
† Georgia Institute of Technology, Atlanta, Ga. 30332

of triethylaluminum in 100 ml. of diethyl ether (Note 3) is then added dropwise through the addition funnel over a 1–2 hour period such that the temperature of the mixture remains below +5°. During the course of the reaction, the color of the mixture changes from light green to brick-red and a reddish-brown crystalline solid forms. When the addition is completed, the mixture is stirred for half an hour at 5–10° and then at 25° for 1–2 hour. The precipitate is collected (Note 4) and washed with several 50-ml. portions of diethyl ether (Note 5). The crude product is purified by extraction at 60° with 400–500 ml. of benzene containing 40 g. of triphenylphosphine. To the dark red, filtered extract is added 200 ml. of *n*-heptane, and the solution is concentrated to *ca.* 300 ml. on a rotary evaporator. An additional 200 ml. of *n*-heptane is added, and the precipitated product is collected and washed with 200 ml. of *n*-heptane and two 80-ml. portions of diethyl ether. This purification is repeated (Note 6), and the reddish-brown crystalline solid dried at 90°/0.1 μ/16 hours to give about 50 g. (55%) of [{(C$_6$H$_5$)$_3$P}$_4$Ni], m.p. 123–128°. *Anal.* Calcd. for C$_{72}$H$_{60}$NiP$_4$: C, 78.3; H, 5.4; Ni, 5.3; P, 11.1. Found: C, 78.7; H, 5.6; Ni, 5.5; P, 10.7.

The complex rapidly decomposes upon exposure to air either as a solid or in solution. It is very soluble in benzene, toluene, and tetrahydrofuran, slightly soluble in diethyl ether, and only very slightly soluble in *n*-heptane and ethanol.

Notes

1. For discussion of general manipulative techniques useful in this work, see D. F. Shriver, "The Manipulation of Air-sensitive Compounds," chap. 7, McGraw-Hill Book Company, New York, 1969.

2. The use of hydrated bis(2,4-pentanedionato)nickel decreases the yield to *ca.* 25%. Ni(C$_5$H$_7$O$_2$)$_2$·xH$_2$O was dehydrated by heating at 100°/0.1 mm./2 hours followed by recrys-

tallization from toluene–heptane and drying at $85°/0.1$ $\mu/16$ hours.

3. Triethylaluminum reacts violently with water and inflames in air. Care must also be taken in preparing the diethyl ether solution since an exothermic reaction occurs; the ether solution is conveniently transferred to the addition funnel via a hypodermic syringe.

4. The filtration is most conveniently performed in a glove box, but various types of glass apparatus designed for the purpose of filtering in an inert atmosphere[3,4] may also be used.

5. The filtrate obtained after collecting the crude product contains highly reactive alkyl aluminum compounds, and contact with water should be avoided. The alkyl aluminum compounds may be decomposed by the dropwise addition of 200 ml. of ethanol to the cooled solution followed by the cautious addition of water.

6. The second recrystallization ensures the removal of a bright orange-red impurity which is formed in low yield.

References

1. G. Wilke, E. W. Müller, and M. Kröner, *Angew. Chem.*, **73**, 33 (1961); German Patent 1,191,375.
2. R. Ugo, *Coord. Chem. Rev.*, **3**, 319 (1968).
3. John J. Eisch and R. Bruce King (eds.), "Organometallic Chemistry," Vol. 1, p. 55, Academic Press Inc., New York, 1965.
4. W. L. Jolly, *Inorganic Syntheses*, **11**, 116 (1968).

25. CARBONYLHYDRIDOTRIS(TRIPHENYL-PHOSPHINE)IRIDIUM(I)

Carbonylhydridotris(triphenylphosphine)iridium(I) has been prepared by the reaction of $[Ir\{P(C_6H_5)_3\}_2(CO)Cl]$ with N_2H_4[1] or $NaBH_4$[2,3] in the presence of excess triphenylphosphine, by treating an alkaline ethanolic suspension of iridium tetraiodide[3] and the phosphine with carbon monoxide, or by the addition of triphenylphosphine to $[Ir\{P(C_6H_5)_3\}_2(CO)H]$.[4] The first procedure described here is a relatively small-scale one (which can, however, be scaled up by a factor of as much as 4) which begins with $[Ir\{P(C_6H_5)_3\}_2(CO)Cl]$. The second procedure, which is presented for a larger scale (but can be scaled down), uses commonly available reactants, and although $[Ir\{P-(C_6H_5)_3\}_2(CO)Cl]$ is prepared as an intermediate, it is not isolated. The reaction sequence includes the reaction of hydrated iridium trichloride, lithium chloride, and carbon monoxide to form an intermediate carbonylchloroiridate(I) salt,* reaction of this solution with triphenylphosphine to form $[Ir\{P(C_6H_5)_3\}_2-(CO)Cl]$,[7] and treatment of the mixture with sodium tetrahydroborate to produce $[\{(C_6H_5)_3P\}_3Ir(CO)H]$. The procedure may be completed in a 24-hour period.

METHOD A

$$[Ir\{P(C_6H_5)_3\}_2(CO)Cl] + P(C_6H_5)_3 + NaBH_4 \rightarrow$$
$$[Ir\{P(C_6H_5)_3\}_3(CO)H]$$

Submitted by G. WILKINSON†
Checked by R. A. SCHUNN‡ and G. L. HENRY‡

Procedure

Absolute ethanol (250 ml.) in a 500-ml. conical flask is deaerated by a stream of nitrogen. Triphenylphosphine (4.5 g.) and *trans*-$[IrCl(CO)(PPh_3)_2]$ (3.24 g.) are added and the solution heated to boiling on a hot plate with stirring by magnetic stirrer. A filtered solution of sodium tetrahydroborate (3 g.) in ethanol

---

* The precise nature of this intermediate is unknown but is most likely $Li[Ir(CO)_2Cl_2]$ by analogy with $K[Ir(CO)_2Br_2]$[5] and $[(C_6H_5)_4As][Ir(CO)_2I_2]$.[6]

† Imperial College of Science and Technology, London, S.W.7, England.

‡ Central Research Department, Experimental Station, E. I. du Pont de Nemours & Company, Wilmington, Del. 19898.

(50 ml.) is added slowly and the solution boiled for about 15 minutes. The product, which is insoluble in ethanol, is collected by filtration of the hot solution through a sintered-glass funnel under nitrogen. The crystals are washed with five portions each of 20 ml. ethanol and dried in vacuum. Yield is 3.4 g. (80%). The compound can be recrystallized from hot toluene as the toluene adduct or from hot benzene by addition of ethanol.

METHOD B

$$IrCl_3 \cdot 3H_2O + 3CO + LiCl \rightarrow$$
$$Li[Ir(CO)_2Cl_2] + 2HCl + CO_2 + 2H_2O$$
$$Li[Ir(CO)_2Cl_2] + 2(C_6H_5)_3P \rightarrow$$
$$[Ir\{(C_6H_5)_3P\}_2(CO)Cl] + CO + LiCl$$
$$[Ir\{(C_6H_5)_3P\}_2(CO)Cl] + (C_6H_5)_3P + NaBH_4$$
$$+ 3CH_3OCH_2CH_2OH \rightarrow [Ir\{(C_6H_5)_3P\}_3(CO)H]$$
$$+ NaCl + B(OCH_2CH_2OCH_3)_3 + 3H_2$$

Submitted by R. A. SCHUNN* and W. G. PEET*
Checked by H. SMITH† and G. WILKINSON†

Procedure

■ *Caution.* *Carbon monoxide is a highly toxic, colorless, and odorless gas and the reaction should be performed only in an efficient fume hood.*

The reaction is performed in a dry nitrogen atmosphere, but no precautions need be taken during the purification steps. A 1-l., three-necked flask is equipped with a magnetic stirrer and a Glass Col heating mantle. One neck of the flask is fitted with a gas inlet tube which will protrude beneath the surface of the 300 ml. of solvent to be used and is attached to a source of carbon monoxide. The second neck of the flask is fitted with an efficient reflux condenser topped with a T tube, one side of

* Central Research Department, Experimental Station, E. I. du Pont de Nemours & Company, Wilmington, Del. 19898.
† Imperial College of Science and Technology, London, S.W.7, England.

which is attached to a source of dry nitrogen and the other side to a silicon oil bubbler. The flask is purged thoroughly with nitrogen and charged with 14.1 g. (0.04 mole) of iridium trichloride trihydrate, 7.6 g. (0.18 mole) of lithium chloride, and 300 ml. of 2-methoxyethanol. The final neck of the flask is stoppered, and the mixture is heated to the boiling point. Stirring will not be efficient until the mixture becomes hot. When the mixture is refluxing, a slow stream of carbon monoxide is passed through the solution for 16 hours (overnight) to produce a clear, brown-yellow solution.*

The solution is cooled to 25°, the gas inlet tube is removed, and 31.5 g. (0.12 mole) of triphenylphosphine is added in portions while maintaining a flow of nitrogen through the flask. Gas (CO) is vigorously evolved and a yellow, crystalline precipitate of $[\{(C_6H_5)_3P\}_2Ir(CO)Cl]$ is formed.† The flask is stoppered, and the mixture is refluxed for half an hour to ensure complete reaction. The yellow slurry is cooled to 25°, and an additional 31.5 g. (0.12 mole) of the phosphine is added. With the nitrogen flow maintained, 4.5 g. (0.12 mole) of sodium tetrahydroborate is added in small portions through one neck of the flask, whereupon gas (H_2) is vigorously evolved. The resulting yellow slurry is stirred at 25° for 0.5 hour and then refluxed gently for 0.5 hour.

The hot mixture is filtered in air, and the yellow solid is washed well with 2-methoxyethanol. The crude product (which may contain some grey contaminant, presumably iridium metal) is extracted with 400 ml. of hot toluene, and the filtered yellow extract is concentrated under vacuum on a rotary evaporator to about 100 ml. The resulting yellow, crystalline solid

---

* If a portion of the reaction mixture is withdrawn after 5 hours and treated with excess triphenylphosphine, impure $[\{(C_6H_5)_3P\}_2Ir(CO)Cl]$ is produced (as judged by infrared spectra); a similar test after 16 hours indicated the formation of pure $[\{(C_6H_5)_3P\}_2Ir(CO)Cl]$.

† This procedure may be utilized for the synthesis of various $L_2Ir(CO)Cl$ complexes[7] although the products may be contaminated with $L_2IrHCl_2(CO)$ derivatives[13] in some cases.

is collected, washed with a little cold toluene, and dried at $75°/0.01$ μ for 4 hours to give 35 g. (80%) of $[\{(C_6H_5)_3P\}_3$-$Ir(CO)H]\cdot C_6H_5CH_3$. *Anal.* Calcd. for $C_{62}H_{54}IrOP_3$: C, 67.8; H, 5.0; P, 8.4. Found: C, 68.1; H, 5.4; P, 8.1.

Properties

The complex $[\{(C_6H_5)_3P\}_3Ir(CO)H]\cdot C_6H_5CH_3$, is a yellow, crystalline solid; m.p. (open capillary) darkens at 165–170°, decomp. >185°. It is stable to storage in air over a period of several months. It is soluble in benzene, toluene, tetrahydrofuran, chloroform, and dichloromethane and insoluble in alcohols, water, and aliphatic hydrocarbons. The unsolvated complex may be obtained in a low-melting form by crystallization from chloroform–hexane (m.p. 145°)[4] and in a high-melting form (m.p. 161°) by recrystallization from benzene–ethanol.[4]

The infrared spectrum (C_6H_6 solution) shows ν_{Ir-H} at 2070 cm.$^{-1}$ and ν_{CO} at 1930 cm.$^{-1}$. The ^1H n.m.r. spectrum in CDCl$_3$ shows the aromatic protons at 3.0τ and the CH$_3$ protons at 7.66τ in the ratio of 50:3 as calculated for 1 mole of toluene in the complex; the Ir—H resonance is observed at 20.7τ as a quartet, $J_{P-H} = 22$ Hz. The unsolvated complex has been shown to be isomorphous with the rhodium analog which has a trigonal bipyramidal structure with the phosphine ligands at equatorial positions.[8]

The complex is active as a hydrogenation catalyst for ethylene and acetylene[9] and as an isomerization catalyst for 1-butene.[10] It reacts with CO to produce $[\{(C_6H_5)_3P\}_2$-$Ir(CO)_2H]$,[3] with triphenyl phosphite to give $[\{(C_6H_5O)_3P\}_3$-$Ir(CO)H]$,[11] and with $(C_6H_5)_2PCH_2CH_2P(C_6H_5)_2$ to give $[\{(C_6H_5)_2PCH_2CH_2P(C_6H_5)_2\}_2IrH]$.[10] Reaction with strong acids produces the $[\{(C_6H_5)_3P\}_3IrH_2(CO)]^+$ cation,[4,12] whereas the deuteride $[\{(C_6H_5)_3P\}_3Ir(CO)D]$ is formed by reaction with D_2.[9]

References

1. S. S. Bath and L. Vaska, *J. Am. Chem. Soc.*, **85**, 3500 (1963).
2. M. Angoletta and G. Caglio, *Rend Ist. Lombardo Sci. Lettere A*, **97**, 823 (1963).
3. G. Yagupsky and G. Wilkinson, *J. Chem. Soc.* (A), **1969**, 725.
4. L. Malatesta, G. Caglio, and M. Angoletta, *ibid.*, **1965**, 6974.
5. L. Malatesta and F. Canziani, *J. Inorg. Nucl. Chem.*, **19**, 81 (1961).
6. L. Malatesta, L. Naldini, and F. Cariati, *J. Chem. Soc.*, **1964**, 961.
7. L. Chatt, N. P. Johnson, and B. L. Shaw, *ibid.* (A), **1967**, 604.
8. S. J. LaPlaca and J. A. Ibers, *J. Am. Chem. Soc.*, **85**, 3501 (1963).
9. L. Vaska, *Inorg. Nucl. Chem. Letters*, **1**, 89 (1965).
10. R. A. Schunn, *Inorg. Chem.*, **9**, 2567 (1970).
11. J. J. Levison and S. D. Robinson, *Chem. Commun.*, **1968**, 1405.
12. L. Vaska, *ibid.*, **1966**, 614.
13. A. J. Demming and B. L. Shaw, *J. Chem. Soc.* (A), **1968**, 1887.

26. CHLOROHYDRIDOTRIS(TRIPHENYLPHOSPHINE)-RUTHENIUM(II)

$$\{(C_6H_5)_3P\}_4RuCl_2 + (C_2H_5)_3N + H_2 \rightarrow$$
$$[\{(C_6H_5)_3P\}_3RuHCl] + (C_2H_5)_3NHCl + (C_6H_5)_3P$$

Submitted by R. A. SCHUNN* and E. R. WONCHOBA*
Checked by G. WILKINSON†

Chlorohydridotris(triphenylphosphine)ruthenium(II) has been prepared by the reaction of $[\{(C_6H_5)_3P\}_3RuCl_2]$ with triethylamine and hydrogen at 25° for 16 hours and pressures of 1 or 120 atmospheres[1] or by reaction of $[\{(C_6H_5)_3P\}_3RuCl_2]$ with sodium tetrahydroborate in a refluxing benzene–H_2O mixture.[1] The procedure described here involves the reaction of $[\{(C_6H_5)_3P\}_4RuCl_2]$, triethylamine, and hydrogen, but may be completed within 2–3 hours.

* Central Research Department, Experimental Station, E. I. du Pont de Nemours & Company, Wilmington, Del. 19898.
† Imperial College of Science and Technology, London, S.W. 7, England.

Procedure

■ *Caution.* *Hydrogen is a highly flammable, explosive gas and the reaction should be conducted in an efficient fume hood.*

The entire procedure is performed in a dry nitrogen atmosphere. A 500-ml., three-necked flask is equipped with a magnetic stirrer and Glass Col heating mantle. One neck of the flask is fitted with a gas inlet tube which will protrude beneath the surface of the 150 ml. of solvent to be used and is attached to a source of dry hydrogen. The second neck of the flask is fitted with an efficient reflux condenser topped with a T tube, one side of which is attached to a source of dry nitrogen and the other to a silicon oil bubbler. The flask is purged thoroughly with nitrogen and charged with 9.5 g. (0.0078 mole) of $[\{(C_6H_5)_3P\}_4RuCl_2]$,[2] 2.4 g. (0.024 mole) of triethylamine, and 150 ml. of deoxygenated toluene. The third neck of the flask is stoppered, and the mixture is stirred and heated to reflux while a slow stream of H_2 is passed through the brown mixture. Hydrogen is passed through the refluxing solution for one hour during which the color changes from brown to purple, and a purple, crystalline solid is formed. The hydrogen flow is stopped, and the mixture is cooled in ice-water for one hour to ensure complete crystallization. The mixture is then filtered in an inert atmosphere (see Note) and the purple, crystalline solid is washed with five 50-ml. portions of deoxygenated ethanol and two 25-ml. portions of deoxygenated toluene. The solid is dried at $25°/0.01\ \mu$ for one hour to give 7.23 g. (90%) of $[\{(C_6H_5)_3P\}_3RuHCl]\cdot C_6H_5CH_3$ m.p. decomp. $>150°$. The product obtained in this manner is sufficiently pure for most uses but may be recrystallized from hot toluene ($\sim$1 g./160 ml.). *Anal.* Calcd. for $C_{61}H_{54}ClP_3Ru$: C, 72.2; H, 5.4; Cl, 3.5; P, 9.1. Found: C, 72.0; H, 5.6; Cl, 3.7; P, 9.4.

Properties

The complex $[\{(C_6H_5)_3P\}_3RuHCl]\cdot C_6H_5CH_3$, is a purple, crystalline solid that decomposes after several hours of exposure to air; solutions of the complex rapidly turn green upon air exposure. It is moderately soluble in chloroform (*ca.* 10^{-3} *M* when saturated[1]) and methylene chloride (*ca.* 10^{-1} *M* when saturated[1]), slightly soluble in benzene, toluene, and tetrahydrofuran, and insoluble in diethyl ether, ethanol, and saturated hydrocarbons.

The infrared spectrum (Nujol mull) shows ν_{RuH} at 2020 cm.$^{-1}$. The ^{1}H n.m.r. spectrum in CD_2Cl_2 shows the aromatic protons at 2.90τ and the CH_3 protons at 7.66τ in the ratio of 51:3 in accord with the presence of 1 mole of toluene in the complex; the Ru—H resonance is observed at 27.75τ as a quartet, $J_{P-H} =$ 26 Hz. The benzene solvate has been shown by x-ray crystallography[3] to have a severely distorted trigonal bipyramidal structure with the triphenylphosphine ligands in equatorial positions.

The complex is an extremely active hydrogenation catalyst for terminal olefins;[1] it is much less effective for the isomerization of olefins.[1] Treatment of the complex with D_2 results in the deuteration of the *ortho*-phenyl positions of the phosphine ligands[1,4] as well as the Ru—D bond.[1,4] It also serves as a catalyst for the preparation of selectively *ortho*-phenyl-deuterated triphenylphosphine.[4]

The complex reacts with triethylaluminum and nitrogen to give $[\{(C_6H_5)_3P\}_3Ru(N_2)H_2]$[5] and with triphenyl phosphite to give the *ortho*-phenyl-bonded phosphite complex $[\{(C_6H_5O)_3P\}_3-\{(C_6H_5O)_2(C_6H_4O)P\}RuCl]$[4,7]. It also reacts with 2,5-norbornadiene and bipyridyl (2,2′-bipyridine) to give $[\{(C_6H_5)_3P\}_2-RuHCl(C_7H_8)]$ and $[\{(C_6H_5)_3P\}_2RuHCl(bipy)]_2$,[1] respectively.

Note

The filtration is most conveniently performed in a glove box, but various types of glass apparatus designed for the purpose of filtering in an inert atmosphere[7,8] may also be used.

References

1. P. S. Hallman, B. R. McGarvey, and G. Wilkinson, *J. Chem. Soc.* (A), **1968**, 3143.
2. T. A. Stephenson and G. Wilkinson, *J. Inorg. Nucl. Chem.*, **28**, 1945 (1966).
3. A. C. Skapski and P. R. Troughton, *Chem. Commun.*, **1968**, 1230.
4. G. W. Parshall, W. H. Knoth, and R. A. Schunn, *J. Am. Chem. Soc.*, **91**, 4990 (1969).
5. W. H. Knoth, *ibid.*, **90**, 7172 (1968).
6. W. H. Knoth and R. A. Schunn, *ibid.*, **91**, 2400 (1969).
7. John J. Eisch and R. Bruce King (eds.), "Organometallic Chemistry," Vol. 1, p. 55, Academic Press Inc., New York, 1965.
8. W. L. Jolly, *Inorganic Syntheses*, **11**, 116 (1968).

Chapter Five

BINARY COMPOUNDS OF THE TRANSITION METALS

27. SINGLE CRYSTALS OF TRANSITION-METAL DIOXIDES

Submitted by D. B. ROGERS,* S. R. BUTLER,*† and R. D. SHANNON*
Checked by A. WOLD‡ and R. KERSHAW‡

Most of the dioxides of the transition metals crystallize in structural types that are closely related to that of the rutile form of titanium dioxide. The series is notable for the wide variety of physical properties and modifications in structure that occur as functions of d-electron number.[1] Electrical transport properties range from insulating to metallic; magnetic properties from Pauli paramagnetic to ferromagnetic. Titanium dioxide has a high index of refraction and is useful as a white pigment and as a dielectric; chromium dioxide is ferromagnetic and has found application in magnetic recording tapes; vanadium dioxide exhibits electrical switching at 68° as conductivity type changes reversibly from semiconducting to metallic. Many of the

* Central Research Department, Experimental Station, E. I. du Pont de Nemours & Company, Wilmington, Del. 19898.

† Present address: Department of Metallurgy and Materials Science, Lehigh University, Bethlehem, Pa. 18015.

‡ Brown University, Providence, R.I. 02912.

heavier dioxides are remarkably good electrical conductors. De Haas-Van Alphen oscillations recently observed[2] in crystals of the dioxides, RuO_2, IrO_2, and OsO_2 were the first such observations on oxidic compounds and provide strong evidence for the formation in these oxides of a primarily *d*-character conduction band that possesses a discrete Fermi surface.

Several high-temperature procedures have been described in the literature for the preparation of the transition-metal dioxides. Direct oxidation of the metals, lower oxides, chlorides, or nitrate precursors provides a convenient route to the dioxides of several metals: Ti, Mn, Ru, Rh, Os, Ir, and Pt.[1,3-5] (Syntheses of the rutile forms of rhodium and platinum dioxides by direct oxidation requires application of high pressures.[5]) Reduction of higher oxides is the most common method of synthesis for these dioxides: VO_2, NbO_2, MoO_2, WO_2, and β-ReO_2.[4,6-8] Stoichiometry in these reactions is most readily controlled by use of the respective metal or a lower oxide as reductant. Chromium dioxide is normally synthesized by hydrothermal reduction of the trioxide.[9]

Except for platinum and rhodium, which have low thermal stabilities, and for technetium, single-crystal growth has been achieved for all the transition-metal dioxides with rutile-related structures. Techniques include flame fusion,[10] electrolytic or thermal reduction from fused salts,[11,12] chemical transport,[13] and extremely high-pressure or hydrothermal recrystallization.[1,14,15] Of these, chemical transport and hydrothermal procedures have most general applicability and appear to lead to products of highest purity. Chemical transport is preferred because it is convenient and utilizes relatively simple laboratory equipment. This technique has been found to be applicable for the crystal growth of titanium dioxide using hydrogen chloride[13,16] or titanium tetrachloride[13] as transporting agents of ruthenium, osmium, and iridium using oxygen,[13] and of niobium, tungsten, and rhenium dioxides using iodine.[1,17] Tungsten dioxide has also been grown by the chemical transport of tungsten trioxide in a reducing hydrogen–water stream.[18]   It seems likely that this rather general technique would also be useful in the growth of crystals of molybdenum and technetium dioxides, perhaps using iodine as the transporting agent, in view of the similar chemistries of these entities to those of tungsten and rhenium, respectively.

Procedures are given here for the crystal growth of the following dioxides: RuO_2, IrO_2, OsO_2, β-ReO_2, and WO_2, using chemical transport techniques.   The procedures described for the dioxides of ruthenium, iridium, and osmium are elaborations of those previously given by Schäfer et al.[13,19,20]

A. RUTHENIUM AND IRIDIUM DIOXIDES

$$RuO_2(s) + \tfrac{1}{2}O_2(g) \rightleftarrows RuO_3(g)$$
$$RuO_2(s) + O_2(g) \rightleftarrows RuO_4(g)$$
$$IrO_2(s) + \tfrac{1}{2}O_2(g) \rightleftarrows IrO_3(g)$$

Procedure

■ *Caution. The volatile higher oxides of ruthenium and iridium are highly toxic. The gas trains used in this procedure should be vented into an efficient fume hood.*   Ruthenium and iridium dioxides can be grown by essentially identical procedures.   These procedures take advantage of chemical transport from a hotter temperature (T_2) to a cooler one (T_1) via volatile higher oxides.   Starting reagents are the polycrystalline dioxides prepared by direct oxidation of ruthenium* and iridium† metal sponges.   A silica boat containing about 5 g. of the respective metal‡ is placed in a silica combustion tube that is fitted via lubricated standard-taper joints to an inlet tube for dry oxygen and an exit tube leading to a water bubbler at the opposite end.   The tube is then placed in a horizontal tube furnace and heated

---

* Engelhard Industries, Newark, N.J. 07114.
† United Mineral and Chemical Company, New York, N.Y. 10013.
‡ The checkers used 2 g. of the relevant metal.

in a slow stream of dry oxygen at 1000° for 24 hours.   During
this process, partial volatilization of the metal oxides will occur
as evidenced by the formation of small (*ca.* $\frac{1}{2}$–1-mm.) crystals
on the downstream end of the silica boat and inner wall of the
combustion tube.   However, maximum crystal size and quality
require more careful control of transport conditions.   The
yield of polycrystalline dioxide remaining in the boat is about
90–95%.   In the case of iridium, oxygen diffusion in the solid
is slow, and the process does not give the dioxide quantitatively.
This fact is not important for the subsequent conversion of
the nonstoichiometric, polycrystalline product to stoichiometric
crystals.

Single-crystal Growth

The apparatus used for crystal growth is shown in Fig. 4 and
essentially consists of an oxygen flow system that permits moni-
toring of flow rate and a gradient furnace that permits careful
control of temperature gradients in the region of growth.   The
growth portion of the flow system is fabricated of fused silica
and consists of an outer combustion tube (130 cm. in length
and 18 mm. i.d.) and an inner tube (60 cm. in length and
15 mm. i.d.) that is split into two semicylindrical halves along
its length.   The function of the split inner tube is to facilitate
removal of the product crystals, which grow on the inner wall

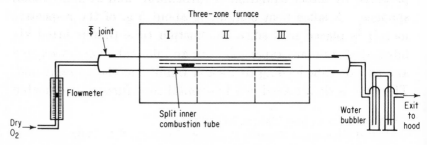

Fig. 4.   Apparatus for crystal growth by chemical transport.

of this tube. The outer combustion tube is equipped at both ends with standard-taper joints that connect on the inlet end to a monitored source of dry oxygen and on the exit end to a water bubbler and final ventilation in a fume hood. The gradient furnace can be of variable design, but should provide a source region at a temperature of 1190° and a growth region at 1090°, with a gradient decreasing from about 8°/cm. in the source region to about 1°/cm. in the growth region. In the apparatus of Fig. 4, these conditions are achieved using a three-zone furnace. Zone I is the source region of highest temperature (1190°). Crystals of maximum quality grow in zone II, where a shallow gradient (*ca.* 1°/cm.) can be maintained over several centimeters at 1090° using a buffer zone (III) set at a slightly higher temperature (*ca.* 1110°).

The combustion tubes are thoroughly washed in distilled water and dried before use. About 2 g.* of ruthenium or iridium dioxide is placed in a $2\frac{3}{4}$-in. boat of recrystallized alumina, and the sample is positioned in zone I of the apparatus, as shown in Fig. 4. Transport is then accomplished with a source-region temperature (T_2) of 1190° and a growth-region temperature (T_1) of 1090° under a stream of oxygen flowing at a rate of about 15–20 cc./minute. Transport is complete in about 15 days, and the operation is terminated by decreasing the flow rate of oxygen to about 2 cc./minute and turning off all furnace power. When the furnace has cooled to room temperature (usually overnight), the split, inner combustion tube is removed and the semicylindrical halves are separated. Crystals will be found all along the downstream length of the tube; however, those of maximum size and quality occur in that portion of the tube exposed to the shallow gradient at 1090°. Generally, these crystals can be removed by gentle tapping of the tube walls.

* The checkers used about 1 g. of the dioxide and carried out the transport for 4 days. The resulting crystals were smaller than those reported by the authors, but had the same properties.

Properties

Ruthenium dioxide is blue-black, and crystals formed in the growth region of the apparatus described above are tabular and about 3–4 mm. in length. Iridium dioxide is somewhat darker (almost black), the normal crystal habit is needlelike, and the crystals are smaller than those of ruthenium dioxide. X-ray powder diffraction patterns taken on ground samples of the crystals can be indexed on the basis of tetragonal unit cells for both dioxides with $a_0 = 4.4906$ A., $c_0 = 3.1064$ A. for RuO_2, and $a_0 = 4.4990$ A., $c_0 = 3.1546$ A. for IrO_2. Both dioxides are Pauli paramagnetic and exhibit metallic conductivity ($\rho_{300°K} \approx 4 \times 10^{-5}$ Ω-cm.). *Anal.* Weight percent oxygen calcd. for RuO_2: 24.05. Found: 24.24. Calcd. for IrO_2: 14.27. Found: 14.40.

B. OSMIUM DIOXIDE

$$3Os + 2NaClO_3 \rightarrow 3OsO_2 + 2NaCl$$
$$OsO_2(s) + O_2(g) \rightleftarrows OsO_4(g)$$

Procedure

■ *Caution. The tetraoxide of osmium, which is involved in the transport process, is more stable, volatile, and toxic than that of ruthenium. It is recommended that the operations described here, which involve the handling of the metal and the oxide at elevated temperatures, be carried out in an efficient fume hood.* Both the preparation of polycrystalline osmium dioxide reagent and its subsequent growth into single-crystal form are conveniently carried out in a single reaction ampul. The ampul is fabricated from a 25-cm. length of 13-mm.-i.d. silica tubing that has been closed at one end. After the tube has been thoroughly washed in distilled water and dried, 0.15 g. of osmium metal powder* and 0.065 g. of sodium chlorate (about 10% excess

* Electronic Space Products, Inc., Los Angeles, Calif. 90035.

over the amount needed for complete conversion of the osmium to the dioxide) are added through a long-stemmed funnel and the tube is connected via rubber tubing to any common vacuum system. When the pressure in the system has been reduced to about 10^{-3} mm., the transport tube is sealed at a length of 15 cm. with an oxy-hydrogen flame. ■ *Caution. The end of the tube containing sodium chlorate and osmium must not be heated during the sealing procedure. This can be prevented by immersing the lower 5 cm. of the tube in a beaker of water or by wrapping such a length in wet asbestos.* The sealed ampul is then slowly heated (at a rate of about 50°/hour) in a muffle furnace to 300° and left overnight. ■ *Caution. Rapid heating at this point must be avoided.* The temperature is then raised to 650° for an additional 3 hours. This treatment results in complete decomposition of the chlorate and formation of golden osmium dioxide powder in one end of the tube. An oxygen pressure of about 0.2 atmosphere results from the excess of sodium chlorate, and about 0.036 g. of sodium chloride is present as the by-product of chlorate decomposition. The oxygen is useful as the transporting agent during subsequent crystal growth; sodium chloride serves no useful function in the reaction, but is not detrimental.

Single-crystal Growth

Chemical transport of osmium dioxide is carried out in a transport furnace wired to provide two independently controllable zones. The transport tube containing oxide, oxygen, and by-product sodium chloride is centered in the two-zone furnace with half of the tube in each of the separate zones. Thermocouples for temperature control are placed at the ends of the tube. It is important for optimum quality and size of product crystals that the growth (empty) zone of the transport tube be free of nucleation sites. To ensure that no microscopic seeds of osmium dioxide are in the growth zone, reverse transport conditions are imposed by heating the growth zone to 960°, while holding the charge at a lower temperature (900°). After several hours of back-transport, the temperatures of the zones are reversed and growth is allowed to proceed with the charge maintained at 960° (T_2) and the growth zone at 900° (T_1). After two days of growth, the furnace is turned off and allowed to cool to room temperature (overnight), and the transport tube is removed and opened.

Properties

Osmium dioxide is golden and crystals formed in the growth zone have an equidimensional habit and are about 2 mm. across a polyhedral face. X-ray powder diffraction patterns taken on powdered crystals can be indexed on the basis of a tetragonal unit cell with a_0 = 4.4968 A. and c_0 = 3.1820 A. The oxide exhibits metallic conductivity ($\rho_{300°K} \approx 6 \times 10^{-5}$ Ω-cm.) and is Pauli paramagnetic. Resistivity ratios ($\rho_{300°K}/\rho_{4.2°K}$) on typical crystals are about 200–300. *Anal.* Weight percent oxygen calcd. for OsO_2: 14.40.   Found: 14.49.

C. TUNGSTEN DIOXIDE AND β-RHENIUM DIOXIDE

$$W + 2WO_3 \rightarrow 3WO_2$$
$$Re + 2ReO_3 \rightarrow 3ReO_2$$

Procedure

The dioxides of tungsten and rhenium are conveniently prepared in powder form for subsequent conversion to single crystals by direct reactions between their respective metal powders and trioxides in a sealed, evacuated system. Both of the metal powders should be freshly reduced in a stream of hydrogen for 3 hours at 1000°. ■ *Caution. Hydrogen forms explosive mixtures with air. The combustion system used for reduction should be thoroughly flushed with nitrogen before admitting hydrogen, and provision for venting the exit gas must be made.*

Tungsten trioxide is predried at 550° for about 2 hours prior to use; rhenium trioxide*† can be used without pretreatment. For the preparation of rhenium dioxide, 2.0 g. of rhenium trioxide and 0.7942 g. of rhenium‡ are added by means of a long-stemmed funnel to a precleaned ampul fabricated by closing one end of a 25-cm. length of 13-mm. i.d. silica tubing. The ampul is then attached *via* rubber tubing to a common vacuum system, the pressure is reduced to about 10^{-3} mm., and finally, the tube is sealed at a length of about 10 cm. using an oxy-hydrogen flame. Reaction inside the sealed ampul to form the dioxide is then accomplished by heating§ the ampul in a muffle furnace at 500° for about 24 hours. The general procedure for the preparation of tungsten dioxide is the same as for rhenium dioxide. However, in this case, 2.0 g. of trioxide is reacted with 0.7932 g. of tungsten at 1100° for 24 hours. Reaction rate is improved by adding 1 or 2 mg. of iodine as a mineralizer to the tungsten reagents prior to evacuation and sealing of the reaction ampul. The products of these reactions are a gray-black powder in the case of rhenium and a golden-brown powder in the case of tungsten.

Single-crystal Growth

Single crystals of tungsten and rhenium dioxides are grown by a procedure that is analogous to that previously described for osmium dioxide, except that iodine is used as the transport agent. The reaction is presumed to involve an oxyiodide and to be of the type:

$$MO_2(s) + I_2(g) \rightleftarrows MO_2I_2(g) \qquad (M = W \text{ or } Re)$$

However, the vapor species involved have not been identified.

---

* Alfa Inorganics, Inc., Beverly, Mass. 01915.

† The checker redried the ReO_3 for 12 hours at 110°.

‡ Electronic Space Products, Inc., Los Angeles, Calif. 90035.

§ The tube was slowly heated to 500° at the rate of 15°/hour in order to prevent explosions.

About 0.5 g. of powdered dioxide and 0.003 g. of iodine are added by means of a long-stemmed funnel to a precleaned, silica ampul fabricated by closing one end of a 25-cm. length of 13-mm.-i.d. tubing. The ampul is evacuated to a pressure of about 10^{-3} mm., sealed at a length of 15 cm., and centered in a two-zone furnace as described in the procedure for growth of OsO_2 crystals. In the case of rhenium dioxide, back-transport to remove stray nuclei is accomplished by heating the growth end of the ampul to 850° while maintaining the charge at 825°; for tungsten dioxide, this is done at temperatures of 1000 and 960° for growth and charge ends, respectively. After back-transport for several hours, the temperatures of the zones are reversed and growth is allowed to proceed. After 3 days of growth, the furnace is turned off, allowed to cool to room temperature, and the transport ampul is removed and opened. Rhenium dioxide crystals should be black; however, the crystals as recovered occasionally have a red-black mottled appearance. This is due to slight surface oxidation with the formation of red trioxide during the cooling procedure. This surface impurity is readily removed by etching the crystals in cold, dilute nitric acid.

Properties

Crystals of tungsten dioxide are golden, and when grown by the procedure described above, they are equidimensional with approximately 2-mm. polyhedral faces. The crystallographic symmetry is monoclinic with unit cell parameters $a_0 = 5.5607$ A., $b_0 = 4.9006$ A., $c_0 = 5.6631$ A., and $\beta = 120.44°$. Tungsten dioxide exhibits metallic conductivity ($\rho_{300°K} \approx 3 \times 10^{-3}$ Ω-cm.). The resistivity ratio ($\rho_{300°K}/\rho_{4.2°K}$) measured for a typical crystal is about 20. *Anal.* Weight percent oxygen calcd. for WO_2: 14.81. Found: 14.89.

Two crystallographic forms of rhenium dioxide are known. When synthesized below 300°, the structural modification (α) is isostructural with tungsten dioxide. Above about 300°, this modification transforms irreversibly to an orthorhombic form (β), and when initial synthesis is carried out at temperatures greater than 300°, the β-modification is invariably recovered. Therefore, the crystals grown by the process described above are β-rhenium dioxide and have an orthorhombic unit cell with $a_0 = 4.809$ A., $b_0 = 5.643$ A., and $c_0 = 4.601$ A. The crystals are black, possess a columnar habit, are about 2–3 mm. in length, and usually are twinned. They exhibit metallic conductivity with $\rho_{300°K} \approx 10^{-4}$ Ω-cm.; however, crystals grown by this process have a relatively low resistivity ratio ($\lesssim 10$). *Anal.* Weight percent oxygen calcd. for ReO_2: 14.67. Found: 14.73.

References

1. D. B. Rogers, R. D. Shannon, A. W. Sleight, and J. L. Gillson, *Inorg. Chem.*, **8**, 841 (1969).
2. S. M. Marcus, S. R. Butler, *Phys. Letters*, **26A**, 518 (1968); S. M. Marcus, paper presented at Am. Phys. Soc. Meeting, Philadelphia, March, 1969.
3. H. Remy and M. Kohn, *Z. Anorg. Allgem. Chem.*, **137**, 381 (1924).
4. G. Brauer, "Handbook of Preparative Inorganic Chemistry," Vol. 2, Academic Press Inc., New York, 1965.
5. R. D. Shannon, *Solid State Commun.*, **6**, 139 (1968).
6. C. Friedheim and M. K. Hoffman, *Ber. Deut. Chem. Ges.*, **35**, 792 (1902).
7. O. Glemser and H. Sauer, *Z. Anorg. Allgem. Chem.*, **252**, 151 (1943).
8. P. Gibart, *Compt. Rend.*, **261**, 1525 (1965).
9. P. Arthur, Jr., U.S. Patent 2,956,955 (1960).
10. J. B. MacChesney and H. J. Guggenheim, *J. Phys. Chem. Solids*, **30**, 225 (1969).
11. D. S. Perloff and A. Wold, "Crystal Growth," p. 361, H. S. Peiser (ed.), Pergamon Press, Ltd., London, 1967.
12. A. M. Vernoux, J. Giordano, and M. Foex, *ibid.*, p. 67.
13. H. Schäfer, "Chemical Transport Reactions," Academic Press Inc., New York, 1964.
14. B. L. Chamberland, *Mat. Res. Bull.*, **2**, 827 (1967).
15. M. L. Harvill and R. Roy, "Crystal Growth," p. 563, H. S. Peiser (ed.), Pergamon Press, Ltd., London, 1967.
16. H. Sainte-Claire Deville, *Ann. Chem.*, **120**, 176 (1861).
17. H. Schäfer and M. Hüesker, *Z. Anorg. Allgem. Chem.*, **317**, 321 (1962).
18. T. Millner and J. Neugebauer, *Nature*, **163**, 601 (1949).
19. H. Schäfer and H. J. Heitland, *Z. Anorg. Allgem. Chem.*, **304**, 249 (1960).
20. H. Schäfer, G. Schneidereit, and W. Gerhardt, *ibid.*, **319**, 327 (1963).

28. MOLYBDENUM(V) FLUORIDE
(*Molybdenum Pentafluoride*)

$$Mo + 5MoF_6 \xrightarrow{60°} 6MoF_5$$

Submitted by T. J. OUELLETTE,* C. T. RATCLIFFE,* D. W. A. SHARP,*
and A. M. STEVEN*
Checked by F. SCHREINER†

The reaction between molybdenum hexacarbonyl and elemental fluoride at $-65°$ results in the formation of Mo_2F_9, which upon thermal degradation produces molybdenum pentafluoride as one of the products.[1]   Other syntheses of molybdenum pentafluoride include the reduction of molybdenum hexafluoride with phosphorus trifluoride,[2] tungsten hexacarbonyl, or molybdenum metal at high temperatures[3] and the oxidation of powdered molybdenum metal with elemental fluoride at 900°.[3]   The present method consists in the reaction of molybdenum hexafluoride with powdered molybdenum metal at 60° and results in the formation of pure molybdenum pentafluoride in yields of 80% and greater.

Procedure

■ *Caution.   Molybdenum hexafluoride is a toxic, highly hygroscopic material, which must be handled in a clean, dry, high-vacuum system.   Molybdenum pentafluoride is also a very hygroscopic compound, which must be handled either in a vacuum line or in an anhydrous, oxygen-free atmosphere.*

A 75-ml. stainless-steel Hoke vessel, which is equipped with a stainless-steel Hoke needle valve (3232 M4S) and a B-10 Monel

* Chemistry Department, University of Glasgow, Glasgow, W.2., Scotland.
† Chemistry Division, Argonne National Laboratory, 9700 South Cass Avenue, Argonne, Ill. 60439.

cone, is attached to a high-vacuum line. The needle valve is opened, and the reaction vessel is degassed several times over a period of 24 hours, using a medium flame with a gas-oxygen torch. The reaction vessel is closed at the needle valve and transferred to a dry-box (P_2O_5). The needle valve is removed and 3.00 g. (0.0313 mole) of high-purity molybdenum powder (Koch-Light Laboratories, Ltd., or Spex Industries Inc., Scotch Plains, N.J. 07076) is added to the reaction vessel. The needle valve is replaced, and the reaction vessel is attached to the high-vacuum line using Kel-F (H200) wax (3M Company), Fig. 5. The molybdenum hexafluoride cylinder (Allied Chemical, Baker-Adamson) is attached to the line with Kel-F (H200) wax via a B-10 cone. Stopcocks (*A–D*) are greased with Kel-F grease. However, Teflon stopcocks (Quickfit, Rotoflo) may also be employed and offer the added advantage of being far more inert to molybdenum hexafluoride.

The reaction vessel is once again degassed in the manner previously described; the glass sections of the line are degassed with a gas–oxygen torch until the glass just glows orange. The

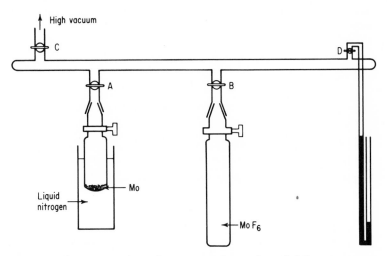

Fig. 5. *Apparatus for the preparation of molybdenum penta-fluoride.*

lower section of the reaction vessel (6 cm.) is cooled to $-196°$ and 0.17 mole of molybdenum hexafluoride condensed.*

The needle valve is closed and the mixture is heated with an oil bath at 60° for 24 hours.

The first portion of molybdenum pentafluoride removed from the reaction vessel will contain as impurities the excess hexafluoride and possibly a small quantity of molybdenum oxyfluoride, $MoOF_4$. The first portion of pentafluoride is separated from the impurities using the apparatus shown in Fig. 6, with all joints sealed using Kel-F wax or using Teflon stopcocks.† The glass collection vessel is carefully degassed as previously described, and then the lower 6 cm. of the reaction vessel is cooled to $-196°$. The molybdenum pentafluoride is removed

---

* The quantity of molybdenum hexafluoride may be measured either in a preweighed vessel or by vapor-pressure measurements in a previously calibrated volume on the line. It was pointed out by the checker that while molybdenum hexafluoride is being handled in the line the stopcock D should normally be kept closed.

† The checker used a copper-to-glass seal to connect the reaction vessel to the glass line in his apparatus for the purification of molybdenum pentafluoride and recommends that the glass line have a downward slope to facilitate the flow of product which was collected in a U-tube cooled in ice-water. A U-tube was not used in the original preparations.

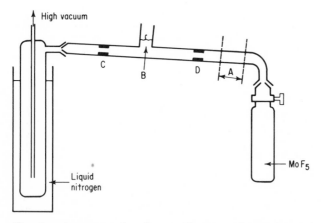

Fig. 6. Apparatus for the purification of molybdenum pentafluoride.

from the reaction vessel under dynamic vacuum. The reaction vessel is opened at the needle valve, and the vessel is slowly warmed to room temperature, which expels all excess hexafluoride. Once all of the hexafluoride has been collected in the −196° trap, the vessel is wrapped with a heating tape, and the temperature is slowly raised to 90–100°. Section A of the collection vessel is wrapped with cotton and kept at −196° by pouring liquid nitrogen over the cotton. The pentafluoride and the more volatile oxyfluoride collects at the −196° section of the vessel as a bright yellow solid with small traces of white oxyfluoride. The withdrawal is continued for a period of 1–2 hours, which results in the collection of 1–1.5 g. of pentafluoride. Care must be employed in heating the reaction vessel for if the temperature exceeds 140°, disproportionation of the pentafluoride takes place:[3]

$$2MoF_5 \rightarrow MoF_6 + MoF_4$$

The reaction vessel is then closed and the cotton removed from the collection vessel. A heating tape is wrapped around the collection vessel to point B (Fig. 6), and the vessel is warmed to 65°. Because of the difference in vapor pressure (at 65°, 4 mm. for the oxytetrafluoride and 2 mm. for the pentafluoride[3]) the oxytetrafluoride will sublime more quickly through the vessel and collect in the trap, while the pure pentafluoride will collect just beyond point B (Fig. 6) as a bright yellow viscous oil. The heating tape is removed and the collection vessel sealed at constrictions C and D.

The pentafluoride collected in the manner described here remains as a viscous oil for a period of 48–72 hours before crystallizing as a bright yellow solid. The remaining product contained in the reaction vessel is of high purity and can either be removed under dynamic vacuum into a vessel similar to that shown in Fig. 6, or stored in the reaction vessel for several weeks without decomposition. Yield is 80% (found by checker, 75%). *Anal.* Calcd. for MoF$_5$: Mo, 50.25. Found: Mo, 50.19.

Properties

Molybdenum pentafluoride is a bright yellow solid with a melting point of 67° and an extrapolated boiling point of 211°.[2] It is very susceptible to hydrolysis; however, it appears quite stable in a dry, stainless-steel vessel and can be handled for short periods in a dry-box (P_2O_5). It is insoluble in most organic solvents but dissolves in dimethyl ether and acetonitrile, giving pale yellow solutions from which addition complexes are obtained.[4]   The solid contains a tetramer, Mo_4F_{20}.

References

1. R. D. Peacock, *Proc. Chem. Soc.*, **1957**, 59.
2. T. A. O'Donnell and D. F. Stewart, *J. Inorg. Nucl. Chem.*, **24**, 309 (1962).
3. A. J. Edwards, R. D. Peacock, and R. W. H. Small, *J. Chem. Soc.*, **1962**, 4486.
4. M. Mercer, T. J. Ouellette, C. T. Ratcliffe, and D. W. A. Sharp, *J. Chem. Soc.* (A), **1969**, 2532.

29. TUNGSTEN(V) CHLORIDE
(*Tungsten Pentachloride*)

$$2WCl_6 + C_2Cl_4 \xrightarrow{h\nu} 2WCl_5 + C_2Cl_6$$

Submitted by E. L. McCANN, III,* and T. M. BROWN*
Checked by C. DJORDJEVIC† and R. E. MORRIS†

Formerly, the preparation of tungsten(V) chloride was accomplished by a hydrogen reduction of tungsten(VI) chloride[1] or more readily by using red phosphorus as the reducing agent.[2] Very pure tungsten(V) chloride can be prepared by disproportionating tungsten(IV) chloride at high temperatures.[3]   It has also been shown that tungsten(VI) chloride decomposes at its boiling point to give tungsten(V) chloride,[4] and there is evidence

* Arizona State University, Tempe, Ariz. 85281.
† The College of William and Mary, Williamsburg, Va. 23185.

that solutions of tungsten(VI) chloride in nonaqueous, nonpolar solvents are unstable, undergoing reduction.[5]

The following procedure[6] takes advantage of the ability of tetrachloroethylene to remove chlorine from solutions of tungsten(VI) chloride and in this manner to bring about a convenient photochemical synthesis of tungsten(V) chloride.

Procedure

The apparatus used for the preparation of tungsten(V) chloride consists of an all-Pyrex glass vessel which can be attached to a standard vacuum line for easy manipulation of the reactants and products. This is shown in Fig. 7.

The reaction vessel consists of a 22-mm. Pyrex tube (closed at one end) with sidearm B terminating in an $\frac{18}{9}$ inner ball joint for loading the vessel. A 13-mm. break seal, C, and an $\frac{18}{9}$ inner ball joint are used for removal of the organic materials after reaction. A small Teflon-coated magnetic stirring bar is placed inside the vessel to aid in thorough mixing of the reactants.

After evacuating and flame-drying the vessel, 4–7 g. of freshly sublimed and powdered tungsten(VI) chloride [cf.: *Inorganic Syntheses*, **9**, 135–136 (1967)] is placed in chamber A of the vessel. This operation is carried out in an inert atmosphere to avoid contamination by oxyhalides. The vessel is then evacuated on the vacuum line, and 25 ml. of previously dried and degassed tetrachloroethylene is vacuum distilled onto the tungsten(VI) chloride. The solution is frozen, and the

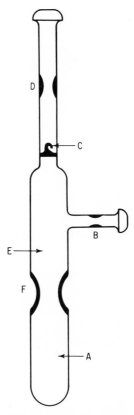

Fig. 7. Reaction vessel for the preparation of tungsten(V) chloride.

vessel is then *carefully* sealed off at *B*. The mixture is then warmed to room temperature, and the sealed vessel is placed in an oil bath at 100°. The reactants are stirred for 24 hours in the presence of a 100-watt light bulb. If the light is too intense, proportionately larger amounts of tungsten(IV) chloride are obtained. The apparatus should be placed behind a safety shield.

During the reaction the solution color changes from a red-brown to a brownish blue-green, and a fine powder appears upon completion of the reaction. The products are then cooled, the apparatus is attached to the vacuum line, and the break seal *C* is broken. The volatile organic materials are then removed by distillation under vacuum, and the vessel is sealed at *D*.

To purify the tungsten(V) chloride, the tube is placed part way in a tube furnace such that a 180°/room-temperature gradient is maintained. The volatile tungsten(V) chloride will then sublime to chamber *E* leaving behind a small amount of nonvolatile, black tungsten(IV) chloride powder. The tube can then be broken at *F* within a dry-box and the product placed in an airtight container for storage.

Yields greater than 90% based on the weight of tungsten(VI) chloride used are obtained depending upon the amount of tungsten(IV) chloride formed. *Anal.* Calcd. for WCl_5: W, 50.91; Cl, 49.09. Found: W, 50.2; Cl, 48.8.

The amount of hexachloroethane formed can be easily determined by gas chromatographic analysis of the organic distillate.

The purity of the tungsten(V) chloride can be determined through inspection of its infrared spectrum. Tungsten oxy-halides and chlorocarbon impurities can be easily identified by their characteristic absorption. An examination of the far-infrared spectrum should show only the characteristic bands of tungsten(V) chloride[7] which can be distinguished from those of other tungsten halide species, although small amounts of such impurities cannot readily be detected in this way.

Properties

The chemical and physical properties of tungsten(V) chloride are well-established.[8-11]   It melts at 248°, boils at 275.6°, and crystallizes in very dark blue-green needles which decompose quickly when exposed to moist air.   The very faint blue-green vapor of tungsten(V) chloride is in contrast to the yellowish-red vapor of tungsten(VI) chloride.   It is only very slightly soluble in nonpolar organic solvents and reacts with most of the polar solvents.

Discussion

With relatively little attention high yields of very pure tungsten(V) chloride can be obtained.   The use of the totally organic reducing medium avoids contamination by species such as phosphorus, and due to the completeness of the reaction, no tungsten(VI) chloride will be found in the product.

Application of this method to other transition-metal systems has been examined with only minor procedural changes.

The quantitative synthesis of molybdenum(IV) chloride has been similarly accomplished in a two-day reaction at 150° in the presence of a 100-watt light bulb.   However, purification cannot be done by sublimation due to the thermal instability of the product.   Thus it is necessary to extract the product with the excess tetrachloroethylene to remove excess molybdenum(V) chloride and hexachloroethane.   The reaction and extraction of the product can be conveniently done in a single vessel as described elsewhere.[12]   This is not as convenient as the method of Larson and Moore[13] but it does avoid contamination by carbonaceous impurities which result in the benzene reduction of molybdenum(V) chloride.

The method has also been examined for the production of other nonvolatile compounds.   In these cases, the purification process involves subliming excess starting materials away from

154   Inorganic Syntheses

the product. For example, greater than 95% yields of tungsten(IV) chloride from tungsten(VI) chloride and 85% yields of niobium(IV) chloride from niobium(V) chloride can be obtained by using a 500-watt light bulb at 150° and a reaction period of 3 days.

References

1. G. Brauer (ed.), "Handbook of Preparative Inorganic Chemistry," Vol. 2, p. 1419, Academic Press Inc., New York, 1965.
2. G. I. Novikov, N. V. Andreeva, and O. G. Ployachenak, *Zh. Neorgan. Khim.*, **6**, 1990 (1961); *Russ. J. Inorg. Chem.*, **6**, 1019 (1961).
3. R. E. McCarley and T. M. Brown, *Inorg. Chem.*, **3**, 1232 (1964).
4. R. Colton and I. B. Tomkins, *Australian J. Chem.*, **19**, 759 (1966).
5. P. M. Boorman, N. N. Greenwood, M. A. Hildon, and R. V. Parish, *Inorg. Nucl. Chem. Letters*, **2**, 377 (1966).
6. T. M. Brown and E. L. McCann, III, *Inorg. Chem.*, **7**, 1227 (1968).
7. E. L. McCann, III, S. T. DeLong, and T. M. Brown, *ibid.* (Submitted for publication).
8. P. M. Boorman, N. N. Greenwood, M. A. Hildon, and H. J. Whitfield, *J. Chem. Soc.*, **1967**, 2017.
9. J. W. Mellor, "Comprehensive Treatise on Inorganic and Theoretical Chemistry," Vol. XI, Longmans, Green & Co., Ltd., London, 1931.
10. P. Pascal (ed.), "Nouveau Traite de Chimie Minerale," Vol. XIV, Masson et Cie, Paris, 1958.
11. N. V. Sidgwick, "The Chemical Elements and Their Compounds," p. 1048, Oxford University Press, London, 1950.
12. T. M. Brown and B. Ruble, *Inorg. Chem.*, **6**, 1336 (1967).
13. M. L. Larson and F. W. Moore, *ibid.*, **3**, 285 (1964).

30. ANHYDROUS NICKEL(II) HALIDES AND THEIR TETRAKIS(ETHANOL) AND 1,2-DIMETHOXYETHANE COMPLEXES

Submitted by LAIRD G. L. WARD*
Checked by J. R. PIPAL†

The synthesis of nickel organic compounds often requires a source of organic-solvent-soluble anhydrous nickel halide.

* The International Nickel Company, Inc., Paul D. Merica Research Laboratory, Sterling Forest, Suffern, N.Y. 10901.
† Massachusetts Institute of Technology, Cambridge, Mass. 02139.

Upon total dehydration,* octahedrally coordinated nickel chloride dihydrate changes to a close-packed cubic structure, or, in the case of hydrated nickel bromide, to a close-packed hexagonal structure.[2] Both of these anhydrous forms dissolve slowly or incompletely in donor solvents. Consequently they have not enjoyed great synthetic utility. Water is consumed but not eliminated during the dehydration of transition-metal halides with 2,2-dimethoxypropane (acetonedimethylacetal),[3] which leaves the transition-metal salt in a mixture of methanol, acetone, and unreacted 2,2-dimethoxypropane. The recent utilization of a complex metal salt, such as tetraalkylammonium tetrahalometallate, as a convenient source of the metal ion in an anhydrous, but hygroscopic, reactive and organic-solvent-soluble form, recognizes this problem.[4] The tris(tetrahydro-furan) complex of chromium(III) chloride[5] affords a useful organic-solvent-soluble anhydrous source of chromium(III). The irreversible, rapid, quantitative hydrolysis of orthoesters[6] and relatively innocuous products has prompted utilization of this reaction as a water-scavenging procedure for nonaqueous solvents.[7]

Crystalline hexakis(ethanol) complexes of perchlorates, tetra-fluoroborates, and nitrates of several metals, Zn, Mg, Mn, Fe, Co, and Ni, have been prepared from the hydrated salts by dehydration with triethyl orthoformate.[8]

Procedures for the preparation of nickel(II) chloride dihy-drate and nickel(II) bromide dihydrate follow. From these, the preparation via the orthoformate ester dehydration route of the anhydrous tetrakis(ethanol) and 1,2-dimethoxyethane complexes is described. The preparation of nickel(II) iodide-bis(1,2-dimethoxyethane) from nickel(II) iodide pentahydrate is also given.

Synthetic applications requiring water-free conditions, such as reactions with Grignard reagents, aryl and alkyl aluminum

---

* Other communications on the preparation of anhydrous metal halides are described in reference 1.

compounds, Group I alkyl and aryl compounds, phosphines and chlorophosphines, alkynes, and weak organic acids render these complexes singularly useful as sources of nickel(II) ion.

A. NICKEL(II) CHLORIDE DIHYDRATE

$$NiCl_2 \cdot 6H_2O \xrightarrow{80°} NiCl_2 \cdot 2H_2O + 4H_2O$$

Procedure

One-inch-thick layers of hydrated nickel chloride are arranged in open Pyrex ovenware dishes which are then stored for 17 hours at 80° in a circulating air oven, continuously purged with dry air. The resultant yellow hygroscopic nickel(II) chloride dihydrate is pulverized in a Waring Blendor and then transferred promptly to securely capped jars. The water content (Karl Fischer titration) varies from 1.88 to 2.2 hydrate.

Properties

Thermogravimetric analysis* of nickel(II) chloride hexahydrate shows that water evolution occurs from ambient temperatures (25°) to 66.6°. The resulting dihydrate is stable up to 133.3°, beyond which temperature further water loss occurs. Differential thermal analysis* shows an endotherm at 53.9° related to the first dehydration step, and a second, strong endotherm at 118.9°, not accompanied by any weight loss, indicates the transformation of an octahedrally coordinated to a close-packed cubic structure.

B. NICKEL(II) BROMIDE DIHYDRATE

$$NiO + 2HBr + H_2O \xrightarrow{80°} NiBr_2 \cdot 2H_2O$$

---

* At a heating rate of 2.5°/minute.

Procedure*

Aqueous hydrobromic acid† (760 g., 4.41 moles, *ca.* 550 ml.) is added to an agitated slurry of black nickel oxide‡ (165 g.) and water (130 ml.) contained in a 2-l. flask, at a rate which maintains the spontaneous exothermic reaction temperature at 80–95°. When the addition of acid is complete, nickel powder§ (3–4 g.) is added to react with the free bromine. The mixture is then boiled briefly, cooled to 26°, and then gravity filtered through a double thickness of fluted filter paper¶ into a 2-l. beaker. The filtrate is vigorously evaporated (1500-watt hot plate) with stirring‖ and an impinging stream of compressed air to a slurry volume of *ca.* 300 ml. and temperature of 130–131°. The product is collected by suction filtration under a covering of rubber dam on a hot (115°), medium-porosity 600-ml. frit. While still hot, the product is promptly transferred to a desiccator where it is cooled *in vacuo* with continued pumping (2 hours/1 mm.).** The product, matted, yellow-brown needles, 240 g. (42%, with 16.9% water of crystallization, Karl Fischer titration), analyzed as a 2.47 hydrate. The water content of

---

* The procedure may be performed in a 12-l. flask using 15 times the quantities indicated here. On a 15× scale, the hydrobromic acid (11,650 g., 67.6 moles, *ca.* 8.4 l.) is added in 500-ml. portions during 4 hours.

† This is 47% w/w 8.56 *M.*; d, 1.38 g./ml., Michigan Chemical Corporation, Box 12, Union, N.J. 07083.

‡ Acid-soluble black nickel oxide, 75.3% nickel, a product of The International Nickel Company, Inc., was used. Properties are described in Ref. 11.

§ Nickel powder type 287, particle size 2.9–3.6 μ; C, 0.05–0.15%; 0, 0.2%, a product of The International Nickel Company, Inc., was used.

¶ The *hot* mixture disintegrates filter paper; the nickel oxide, nickel powder, and fine carbon residues rapidly block a sintered-glass frit or Celite filter bed.

‖ Ace Glass Inc., catalog no. 8256-G, stirring rod; 8258-A $2\frac{1}{2}$-in., paddle-type Teflon agitator blade.

** Needles of the dihydrate form in boiling concentrated aqueous solutions. As the temperature falls during their isolation, the water content increases and can vary from 2.14–2.85 or more moles per mole of nickel bromide.

it was reduced to 14.7% or 2.09 hydrate by drying $\frac{1}{4}$-in.-thick layers at 95° for one hour in a circulating air oven.*

Properties

Thermogravimetric analysis† of a sample of the 5 hydrate shows that water evolution occurs between 34.1° and 89.6°, at which latter temperature a dihydrate has formed. This is stable up to 107°, beyond which temperature the remaining two water molecules are slowly lost. Differential thermal analysis† shows two dehydration endotherms at 36.4 and 132.8° and a structure transformation (octahedrally coordinated to close-packed hexagonal) endotherm at 151.8°.

C. DICHLOROTETRAKIS(ETHANOL)NICKEL(II)

$$NiCl_2 \cdot 2H_2O + 2HC(OC_2H_5)_3 \xrightarrow[\text{reflux}]{C_2H_5OH}$$
$$NiCl_2 \cdot 4C_2H_5OH + xsC_2H_5OH + 2HCOOC_2H_5$$

Procedure‡

A three-necked, 3-l. flask fitted with a reflux condenser and mechanical stirrer is charged with 327 g. (2.0 moles) of nickel(II) chloride 1.88 hydrate, 550 ml. of absolute ethanol, and 558 g. (4.14 moles) of triethyl orthoformate.§ The cream-colored slurry is stirred at the reflux under an atmosphere of nitrogen for 2 hours, after which time the water content of a 1-ml. sample of the now homogeneous reaction mixture is checked with a

---

* Prolonged drying (65 hours) at this temperature will yield a powdery brown hydrated product (2% H_2O) which is insoluble, or only extremely slowly soluble, in donor solvents.

† At a heating rate of 2.5°/minute.

‡ The technique of chemical manipulation under a nitrogen atmosphere, using a Schlenk frit apparatus like that described in Fig. 6 of *Inorganic Syntheses*, **11**, 78 (1968), is further outlined in reference 9. Teflon plug stopcocks, which do not require lubrication, are preferred.

§ Kay-Fries Chemicals, Inc., 360 Lexington Ave., New York, N.Y. 10017.

Karl Fischer titration.* The Karl Fischer analysis should indicate that the solution is completely water-free. If water is present the mixture should be refluxed for an additional hour. Another Karl Fischer analysis should then be performed, and if water continues to persist, an additional 0.1–0.2 mole of triethyl orthoformate should be added and the refluxing continued for an additional hour. Upon concentration of the solution to a volume of 1 l., and then cooling it to 0° under nitrogen to exclude moisture, pastel-green hygroscopic needles of the tetrakis-(ethanol) complex (140 g., 22%†) form. These are collected and carefully dried under nitrogen on a Schlenk frit at 23°.‡ *Anal.* Calcd. for $C_8H_{24}ClO_4Ni$: C, 30.61; H, 7.71; Cl, 22.59; O, 20.39; Ni, 18.70. Found: C, 27.71; H, 6.95; Cl, 22.97; O, 20.82; Ni, 19.09.

Properties

The bis(ethanol) complex described by Ostoff and West[10] is not isolated by this procedure, but evidence for its presence is apparent on thermogravimetric analysis. When heated slowly,§ the tetrakis(ethanol) complex gradually loses ethanol up to 54°. The resulting bis(ethanol) complex then continues to gradually lose ethanol, and the material at 68° has an estimated composition $NiCl_2·1.8C_2H_5OH$. At 90°, a monoethanol complex forms. Differential thermal analysis§ shows three endotherms, all related to volatilization of the ligand: 38.4°, associated with the formation of $NiCl_2·2C_2H_5OH$; 80.9°, associated with the formation of $NiCl_2·C_2H_5OH$; and 89.9°, related to the loss of the remaining ethanol. The tetrakis(ethanol) complex is *very* soluble in the lower alcohols, esters, and monoethers of eth-

* A 2-ml. Insulin B-D Yale Luer-lok multifit hypodermic syringe with a 6-in. 18G needle is used to conveniently transfer the analytical sample.

† The checker obtained a 65% yield working on $\frac{1}{6}$ this scale.

‡ Excessive drying will produce a yellow-orange-colored product with approximately 3 moles of ethanol of crystallization per mole of nickel chloride.

§ At a heating rate of 2.5°/minute.

yleneglycol (Cellosolves) and soluble in tetrahydrofuran. It is not soluble in hydrocarbons, diethyl ether, or methylene chloride.

D. DIBROMOTETRAKIS(ETHANOL)NICKEL(II)

Procedure

Following the procedure described under C, a slurry of 526 g. (2.0 moles) of nickel(II) bromide 2.47 hydrate and 800 g. (5.4 moles) of triethyl orthoformate is stirred at the reflux for 2 hours under 1 atmosphere of nitrogen. The resulting homogeneous solution should at this stage be water-free (Karl Fischer titration of a 1-ml. sample). It is concentrated to a volume of 1 l. and the dull-yellow chunky hygroscopic needles (210 g., 26%*) which slowly form are collected and carefully dried under nitrogen on a Schlenk frit at 23°.† *Anal.* Calcd. for $C_8H_{24}BrO_4Ni$: C, 23.85; H, 6.00; Br, 39.68; O, 15.89; Ni, 14.57. Found: C, 21.22; H, 5.45; Br, 39.95; O, 15.42; Ni, 14.59.

Properties

The tetrakis(ethanol) complex readily changes to a brown tris(ethanol) complex when it is dried at reduced pressure. Thermogravimetric analysis‡ shows ethanol is evolved from the tetrakis(ethanol) complex below 26° and this loss is complete at 113°. Two endotherms are observed on differential thermal analysis‡ during the final 4% loss in weight, at 80 and at 89°.

E. DICHLORO(1,2-DIMETHOXYETHANE)NICKEL(II)

$$NiCl_2 \cdot 2H_2O + 2HC(OC_2H_5)_3 + CH_3OCH_2CH_2OCH_3 \xrightarrow{reflux}$$
$$NiCl_2 \cdot C_4H_{10}O_2 + 4C_2H_5OH + 2HCOOC_2H_5$$

* The checker obtained a 50% yield working on ⅛ this scale.
† Excessive drying will result in the loss of some of the bound ethanol.
‡ At a heating rate of 2.5°/minute.

Procedure

A three-necked, 3-l. flask fitted with a reflux condenser and mechanical stirrer is charged with 331 g. (2.0 moles) of pulverized nickel(II) chloride 1.88 hydrate, 1 l. of 1,2-dimethoxyethane,* and 651 g. (4.4 moles) of triethyl orthoformate. The slurry is stirred rapidly and heated at the reflux under nitrogen for 2 hours, after which time a Karl Fischer water analysis of the supernatant green liquid should show less than 0.04 mg. H_2O/ml. The completed reaction slurry is cooled, and the orange granular solids are collected under nitrogen on a Schlenk frit of 1-l. capacity, rinsed successively with anhydrous 1,2-dimethoxyethane and then pentane, and dried in a nitrogen atmosphere at 26° and 20 cm. Hg. The yield is nearly quantitative. *Anal.* Calcd. for $C_4H_{10}Cl_2O_2Ni$: C, 21.86; H, 4.59; Cl, 32.27; Ni, 26.72. Found: C, 21.59; H, 4.72; Cl, 31.98; Ni, 26.54.

Preparation of Related Compounds. The addition of diglyme or of triglyme† to concentrated anhydrous ethanol solutions of nickel(II) chloride afforded $NiCl_2 \cdot$diglyme and $NiCl_2 \cdot$(triglyme)$_{0.5}$, respectively.

Properties

Whether prepared from a homogeneous (excess ethanol is required) or a nonhomogeneous medium, the orthoformate derived dichloro(1,2-dimethoxyethane)nickel(II), on thermogravimetric analysis (heating at 10°C. min^{-1}) begins to lose its ligand at 91°. The loss proceeds in two stages: the first, 28%

---

* 1,2-Dimethoxyethane (monoglyme) was refluxed over sodium metal until no further reaction occurred with a clean sodium surface. It was distilled from clean sodium; b.p. 83°/775 mm.

† Diglyme, [bis(2-methoxyethyl)ether], and triglyme, [1,2-bis(methoxyethoxy)ethane; 2,5,8,11-tetraoxydodecane], were each refluxed 4 hours over calcium hydride granules, then distilled from calcium hydride at atmospheric pressure; fractions b.p. 161–162°/750 mm. and 222–224°/750 mm., respectively, were collected.

loss, corresponding to 0.7 1,2-dimethoxyethane is complete at 155° and is accompanied by a large endotherm at 152°. The second 12% loss (0.3 1,2-dimethoxyethane) is complete at 287° but without an accompanying endo- or exothermal process.

F. DIBROMO(1,2-DIMETHOXYETHANE)NICKEL(II)

Procedure

The procedure described in Sec. D for the preparation of dibromotetrakis(ethanol)nickel(II) is followed. The anhydrous ethanol solution is evaporated to the stage of incipient crystallization at the boiling point* and then diluted with 1 l. of peroxide-free anhydrous 1,2-dimethoxyethane. The resulting salmon-pink complex is collected under nitrogen in a Schlenk frit of 1-l. capacity, rinsed successively with anhydrous 1,2-dimethoxyethane and then pentane and dried in a current of nitrogen at 23° and 20 cm. Hg. The yield is nearly quantitative.† *Anal.* Calcd. for $C_4H_{10}Br_2O_2Ni$: C, 15.57; H, 3.27; Br, 51.78; Ni, 19.02. Found: C, 15.01; H, 3.43; Br, 52.15; Ni, 18.80.

Preparation of Related Compounds. The addition of diglyme or of triglyme to concentrated anhydrous ethanol solutions of nickel bromide afforded $NiBr_2 \cdot$diglyme and $NiBr_2 \cdot$triglyme, respectively.

Properties

The loss of the one ligand molecule from dibromo(1,2-dimethoxyethane)nickel(II) commences at 114° and is complete

---

* Sufficient ethanol should be present to provide a homogeneous solution at the boiling point. Due to the appreciable solubility of the dibromo(1,2-dimethoxyethane)nickel in ethanol-1,2-dimethoxyethane mixtures, the yield of the salmon-pink complex will be lower if enough ethanol–ethyl formate is not volatilized.

† The checker obtained a 70% yield working on $\frac{1}{8}$ this scale.

at 275°.   This loss is accompanied by one endothermal process
at 164°.

G. BIS(1,2-DIMETHOXYETHANE)DIIODONICKEL(II)

Procedure

Crystalline nickel iodide pentahydrate (100.6 g., 0.25 mole)
is added to stirred triethyl orthoformate (200 g., 1.35 moles).
When the water content of the solution is less than 0.04 mg./ml.
(3 hours), the volatiles are stripped at reduced pressure.   The
black residue is recrystallized from 200 ml. anhydrous, peroxide-
free, hot 1,2-dimethoxyethane in which it is very soluble,
forming a blood red solution.   The solvent is evaporated
slowly at approximately 23° and 20 cm. Hg.   The large (1 mm.)
deep orange-red granular hygroscopic crystals (60 g., 0.12 mole,
48%) are collected under nitrogen in a Schlenk frit, washed with
pentane, and dried in nitrogen at 23° and 20 cm. Hg.   *Anal.*
Calcd. for $C_8H_{20}I_2O_4Ni$: C, 19.50; H, 4.09; I, 51.51; Ni, 11.91.
Found: C, 17.99; H, 4.02; I, 50.11; Ni, 11.62.

Properties

Bis(1,2-dimethoxyethane)diiodonickel(II) begins to lose some
of its dimethoxyethane in a nitrogen atmosphere even at ambient
temperatures.   This sample analyzed for 95% of the two
liganded 1,2-dimethoxyethane molecules which were originally
present.   At 88°, this loss amounted to two more percent; then,
further increasing the temperature, the remaining 1,2-dimeth-
oxyethane is volatilized up to 133°.   The 34% loss (calcd.
36.6%) is accompanied by an intense endotherm at 109°, which
is followed by a smaller endotherm at 135°.

Properties of the 1,2-Dimethoxyethane Complexes.   The degree
of solubility is greatest for the iodide complex and least for the
chloride complexes.   The sensitivity to moisture—that is, the

relative ease with which the ligand is replaced by water—present as moisture in the atmosphere—is greatest for the iodide and least for the chloride. Consequently, it has been found possible, by working quickly, to weigh the chloride and bromide complexes in air and to transfer them to the reaction vessel, without any appreciable contamination with water from the atmosphere.

The complexes are very soluble in methanol, ethanol, butanol, methyl Cellosolve (2-methoxyethanol), and ethyl Cellosolve (2-ethoxyethanol), and to a fair degree, quite soluble in 1,2-dimethoxyethane and di- and triglyme. They are initially quite soluble in tetrahydrofuran, acetone, pyridine, nitromethane, acetonitrile, dimethyl sulfoxide, and N,N-dimethylformamide, but usually precipitation of the nickel halide–solvent complex occurs if attempts are made to prepare moderately concentrated solutions in these solvents. They are only very slightly soluble, or are quite insoluble in dioxane, ethyl ether, hexane, dichloromethane, ethyl acetate, and methyl- and butylcellosolve acetate (2-methoxyethyl and 2-butoxyethyl acetate).

References

1. S. Y. Tyree, Jr., *Inorganic Syntheses*, **4**, 104 (1953); A. R. Pray, *ibid.*, **5**, 153 (1957); E. R. Epperson, S. M. Horner, K. Knox, and S. Y. Tyree, Jr., *ibid.*, **7**, 163 (1963); W. W. Porterfield and S. Y. Tyree, Jr., *ibid.*, **9**, 133 (1967).
2. A. F. Wells, "Structural Inorganic Chemistry," 3d ed., pp. 345, 875, Oxford University Press, London, 1962.
3. J. Starke, *J. Inorg. Nucl. Chem.*, **11**, 77 (1959).
4. R. H. Holm, F. Röhrscheid, and G. W. Everett, Jr., *Inorganic Syntheses*, **11**, 74 (1968).
5. James P. Collman and E. T. Kittleman, *ibid.*, **8**, 150 (1966).
6. R. H. DeWolfe and R. M. Roberts, *J. Am. Chem. Soc.*, **76**, 4379 (1954); J. Hine, Abstracts of 18th National Organic Chemistry Symposium, p. 85, The American Chemical Society, Washington, D.C., 1963.
7. G. Kesslin and R. Bradshaw, *Ind. Eng. Chem., Prod. Res. Develop.*, **5**, 27 (1966).
8. P. W. N. M. van Leeuwen, *Rec. Trav. Chim.*, **86**, 247 (1967).
9. S. Herzog and J. Dehmert, *Z. Chem.*, **4**, 1 (1964).
10. R. C. Ostoff and R. C. West, *J. Am. Chem. Soc.*, **76**, 4732 (1954).
11. A. Illis, G. W. Nowlan, and H. J. Koehler, *Can. Inst. Mining Met. Bull. Transactions*, **73**, 44 (1970).

Chapter Six

HALO COMPLEXES OF SOME METALS

31. TETRAPHENYLARSONIUM
TETRACHLOROVANADATE(III)

$$VCl_3 + 3CH_3CN \rightarrow [VCl_3 \cdot 3CH_3CN]$$
$$[VCl_3 \cdot 3CH_3CN] + [(C_6H_5)_4As]Cl \rightarrow [(C_6H_5)_4As][VCl_4 \cdot 2CH_3CN]$$
$$[(C_6H_5)_4As][VCl_4 \cdot 2CH_3CN] \xrightarrow{100°/10^{-2} \text{ mm.}} [(C_6H_5)_4As][VCl_4] + 2CH_3CN$$

Submitted by A. T. CASEY,* R. J. H. CLARK,† R. S. NYHOLM,† and D. E. SCAIFE‡
Checked by T. E. BOYD,§ W. RHINE,§ and G. STUCKY§

The great majority of compounds of vanadium(III) have octa-hedral or quasioctahedral symmetry about the metal atom. Recently, the preparations of the first tetrahedral vanadium(III) species have been described by the authors.[1,2] The general method of their preparation is exemplified by that of tetra-phenylarsonium tetrachlorovanadate(III) described here.

Procedure

Reagent-grade acetonitrile is dried over molecular sieves for 48 hours and then fractionally distilled from tetraphosphorus

* University of Melbourne, Australia.
† University College, Gower St., London, W.C.1, England.
‡ Division of Mineral Chemistry, C.S.I.R.O., Melbourne, Australia.
§ University of Illinois, Urbana, Ill. 61801.

decaoxide three times. A unit similar to that illustrated in Fig. 8 is flamed out under vacuum and charged with 6.1 g. anhydrous vanadium trichloride (0.039 mole) in an argon-filled dry-box. One-hundred twenty-five milliliters of the purified acetonitrile is distilled from tetraphosphorus decaoxide in a stream of dry argon or nitrogen into the unit, using only the fraction boiling at 82.0–82.5°. The mixture of vanadium trichloride and acetonitrile is warmed at 60° until solution is com-

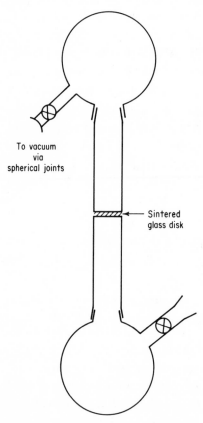

To vacuum
via
spherical joints

Sintered
glass disk

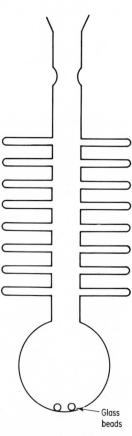

Glass
beads

Fig. 8. Apparatus for preparing derivatives of vanadium trichloride.

Fig. 9. Manifold for storing and taking samples of salts of the tetrahalovanadate(III) ion.

plete; the vanadium(III) solvate has the formula [$VCl_3 \cdot 3CH_3CN$] and is green.

Tetraphenylarsonium chloride (19.0 g., 0.045 mole) is dried *in vacuo* at 100° for 6 hours. The salt is dissolved in 40 ml. dry acetonitrile in the dry-box, the atmosphere of which is kept reasonably free of acetonitrile vapor by using serum-capped containers and transferring the acetonitrile by hypodermic syringe. The solution of tetraphenylarsonium chloride is added to the solution of vanadium trichloride in acetonitrile all at once against a stream of dry argon, the mixture heated to boiling and allowed to cool slowly. Bright yellow crystals of [$(C_6H_5)_4As$][$VCl_4 \cdot 2CH_3CN$] precipitate as the solution cools. These are filtered at the sintered-glass disk by the application of gentle vacuum. Pumping the solid at 20° for 4 hours gives the pure solid. The compound is then transferred in a dry-box to a manifold of the type illustrated in Fig. 9 and pumped at 100° for 6 hours at 0.01 mm. The two molecules of acetonitrile are removed easily under these conditions to yield the bright blue compound [$(C_6H_5)_4As$]VCl_4. Yield is 15.1 g. (59% based on VCl_3). *Anal.* Calcd. for [$(C_6H_5)_4As$][$VCl_4 \cdot 2CH_3CN$]: C, 50.3; H, 3.55; Cl, 21.9; N, 3.0; V, 8.25. Found: C, 51.1; H, 4.0; Cl, 21.5; N, 4.25; V, 7.75. Calcd. for [$(C_6H_5)_4As$]VCl_4: C, 50.0; H, 3.5; Cl, 24.6; V, 8.84. Found: C, 50.0; H, 3.0; Cl, 23.9; V, 9.20.

Properties

In common with most other complexes of vanadium(III), these compounds are highly susceptible to hydrolysis and oxidation. The magnetic moment of [$(C_6H_5)_4As$][VCl_4] is independent of temperature[2] as one would expect for a complex with $^3A_2$ ground term. The intensities and spacings of the x-ray powder pattern[1] are virtually identical with those of [$(C_6H_5)_4As$][$FeCl_4$], the structure of which is known[3] to be a somewhat distorted tetrahedron. It has not been possible to obtain a solution spectrum, since every solvent in which the compound is soluble appears to increase the coordination number. The reflectance spectrum can be assigned on a tetrahedral model.[2]

Using a similar procedure it is possible to prepare $[(C_6H_5)_4As]$-$[VBr_4]$, $[(C_2H_5)_4N][VBr_4]$, and $[(C_2H_5)_4N][VCl_4]$. The first of these is probably tetrahedral as well, but it is so susceptible to solvation that it is difficult to work with even in a carefully dried dry-box. The second compound contains the tetrahedral VBr_4^- ion, but the last (which was mentioned in passing in *Inorganic Syntheses*, volume **11**, page 79)[4] is an octahedral polymer as indicated by both spectral and magnetic measurements.[2]

References

1. R. J. H. Clark, R. S. Nyholm, and D. E. Scaife, *J. Chem. Soc.*, **1966**, 1296.
2. A. T. Casey and R. J. H. Clark, *Inorg. Chem.*, **7**, 1598 (1968).
3. B. Zaslow and R. E. Rundle, *J. Phys. Chem.*, **61**, 490 (1957).
4. R. H. Holm, F. Röhrscheid, and G. W. Everett, *Inorganic Syntheses*, **11**, 79 (1968).

32. TRIS(TETRAETHYLAMMONIUM) NONACHLORODIVANADATE(III)

$$2VCl_3 + 3(C_2H_5)_4NCl \xrightarrow{SOCl_2} [(C_2H_5)_4N]_3[V_2Cl_9]$$

Submitted by A. T. CASEY* and R. J. H. CLARK†
Checked by W. RHINE,‡ T. E. BOYD,‡ and G. STUCKY‡

The $M_2Cl_9^{3-}$ ions where M = vanadium, chromium, or tungsten, are binuclear consisting of two octahedra with shared edges.[1] Corresponding ions in which M = Ti or Mo presumably have the same structure. The $Fe_2Cl_9^{3-}$ ion is known[2] only in associa-

* University of Melbourne, Australia.
† University College, Gower St., London, W.C.1, England.
‡ University of Illinois, Urbana, Ill. 61801.

tion with the cesium ion.   Slight differences in the metal-metal distance in the binuclear anions cause profound changes in their magnetic susceptibilities; for instance, the tungsten compound is diamagnetic while the chromium compound shows evidence of only a slight interaction between the metal atoms.[3]   The vanadium compound is intermediate in this respect.[4]   The procedure used for the preparation of the vanadium compound was first described briefly in 1963[5] and involves the use of thionyl chloride as the solvent.

Procedure

The apparatus is the same as that used in the preparation of $[(C_6H_5)_4As]VCl_4$ (page 166).   After flaming under vacuum the unit is charged, in an argon- or nitrogen-filled dry-box, with 5.2 g. (0.033 mole) of finely ground anhydrous vanadium trichloride.   Thionyl chloride (125 ml.) is added against a stream of dry argon.   Tetraethylammonium chloride (10.0 g., 0.060 mole), dried under vacuum, is dissolved in 25 ml. of thionyl chloride and added to the mixture of vanadium trichloride and thionyl chloride.   The unit is shaken gently for 30 minutes and then allowed to stand for 24 hours with occasional agitation. The deep red solution is filtered through the sintered-glass disk by application of a slight vacuum and the thionyl chloride removed under vacuum (*ca.* 10 mm.) until crystallization begins. A 2:1 mixture of diethyl ether–thionyl chloride (30 ml.) is then added to complete the crystallization.   The sintered-glass filter is changed in the dry-box and the crystals collected by filtration using a slight vacuum.   The crystals are washed with the ether–thionyl chloride mixture three times and then dried at about 50°/0.1 mm.   Yield is 8.7 g. (65% based on VCl_3).   Working on one-fourth this scale the checkers obtained a 44% yield. *Anal.* Calcd. for $[(C_2H_5)_4N]_3V_2Cl_9$: C, 35.51; H, 7.44; Cl, 39.31; N, 5.18; V, 12.55.   Found: C, 34.90; H, 8.26; Cl, 39.31; N, 5.67; V 12.86.

Properties

Tris(tetraethylammonium) nonachlorodivanadate(III) forms red crystals which are very easily hydrolyzed and easily oxidized; thus they are most conveniently stored in a manifold such as that illustrated in Fig. 9.   The solid is placed in the manifold in a dry-box and then sealed off *in vacuo*.   The glass balls help to prevent the solid caking.   When a sample is required it is tapped into one of the sidearms, which is then sealed off.   The spectral[4,5] and magnetic[4,6] properties of the compound are consistent with some vanadium–vanadium interaction within the binuclear anion.

References

1. G. J. Wessel and D. J. W. Ijdo, *Acta Cryst.*, **10**, 466 (1957).
2. R. J. H. Clark and F. B. Taylor, *J. Chem. Soc.* (A), **1967**, 693.
3. A. Earnshaw and J. Lewis, *ibid.*, **1961**, 396.
4. A. T. Casey and R. J. H. Clark, *Inorg. Chem.*, **7**, 1598 (1968).
5. D. M. Adams, J. Chatt, J. M. Davidson, and J. Gerratt, *J. Chem. Soc.*, **1963**, 2189.
6. R. Saillant and R. A. D. Wentworth, *Inorg. Chem.*, **7**, 1606 (1968).

33. AMMONIUM, RUBIDIUM, AND CESIUM SALTS OF THE AQUAPENTACHLOROMOLYBDATE(III) AND HEXACHLOROMOLYBDATE(III) ANIONS

$$Mo_2(O_2CCH_3)_4 + HCl(aq.) + O_2 \rightarrow [MoCl_5(H_2O)]^{2-} + [MoCl_6]^{3-}$$

$$[MoCl_5(H_2O)]^{2-} + Cl^- \xrightarrow[\text{HCl}]{\text{satd. aq.}} [MoCl_6]^{3-} + H_2O$$

$$[MoCl_6]^{3-} + H_2O \xrightarrow{6\ M\ HCl} [MoCl_5(H_2O)]^{2-} + Cl^-$$

Submitted by J. V. BRENCIC* and F. A. COTTON*
Checked by J. A. JAECKER† and R. A. WALTON†

The preparation of the compounds $K_3[MoCl_6]$ and $K_2[MoCl_5(H_2O)]$ is described in an earlier volume in this series.[1]   The

* Massachusetts Institute of Technology, Cambridge, Mass. 02139.
† Purdue University, Lafayette, Ind. 47907.

procedures are relatively arduous, for three reasons. (1) They begin with electrolytic reduction of MoO_3 dissolved in hydrochloric acid—a lengthy and tedious procedure. (2) The law of mass action is not employed efficiently to maximize the desired product. (3) The choice of potassium as the M^I ion is unwise unless there is a special reason for wanting it, since potassium chloride is relatively insoluble in hydrochloric acid and therefore tends to contaminate the products.

We describe here a procedure for preparing the ammonium, rubidium, and cesium salts of $[MoCl_5(H_2O)]^{2-}$ and $[MoCl_6]^{3-}$ in good yields employing tetrakis(acetato)dimolybdenum,[2] which is itself easily prepared[2] from $Mo(CO)_6$. The procedure takes deliberate advantage of the law of mass action, in that the $[MoCl_5(H_2O)]^{2-}$ compounds are precipitated from 6 M hydrochloric acid, whereas the $[MoCl_6]^{3-}$ salts are obtained from saturated aqueous hydrogen chloride. Finally, the potassium cation is avoided.

Procedure

A. AQUAPENTACHLOROMOLYBDATE(III) SALTS

Three grams (7.01 mmoles) of tetrakis(acetato)dimolybdenum and 2.0 g. ammonium chloride (or 5.0 g. RbCl or 6.5 g. CsCl; *ca.* 40 mmoles) are mixed with 50 ml. of 12 M aqueous hydrochloric acid. The slurry is warmed to the boiling point and evaporated to about 10 ml., giving a clear, red solution. This solution is then placed in an ice bath and left for at least 6 hours, during which the $M^I{}_2[MoCl_5(H_2O)]$ precipitates as a brick-red, crystalline solid. This precipitate is collected by filtration, washed with two 20-ml. portions of absolute ethanol and dried in vacuum at 25° for 10 hours. The yields are usually about 75% {e.g., 3.4 g. of $[NH_4]_2[MoCl_5(H_2O)]$}. *Anal.* Calcd. for $[NH_4]_2[MoCl_5(H_2O)]$: N, 8.56; Cl, 54.16; Mo, 29.38. Found: N, 8.57; Cl, 54.3; Mo, 29.2.

■ *Note. The solvent from which the product precipitates is*

*constant-boiling hydrochloric acid, which is 20.2% HCl, by weight,
ca. 6 M and boils at 108.6°/760 mm.*[3]

B. AMMONIUM HEXACHLOROMOLYBDATE(III)

Two and one-half grams of $[NH_4]_2[MoCl_5(H_2O)]$ and 0.5 g.
of ammonium chloride are added to 40 ml. of 12 M hydrochloric
acid in a flask.   The slurry is boiled gently until all solids have
dissolved.   This usually requires about 20 minutes.   The flask
is then removed from the source of heat and a slow stream of
gaseous hydrogen chloride is bubbled through the solution until
its temperature falls to about 50°.   Absolute ethanol (5 ml.) is
then added, and the solution is placed in an ice bath while the
passage of hydrogen chloride is continued.   When the solution
has cooled fully and is saturated with hydrogen chloride as
indicated by the fumes emerging from it, the hydrogen chloride
stream is stopped and the stoppered solution left in the ice bath
for an hour.   Total time to this point is about 2 hours.

The pink crystals can now be collected by filtration in air.
They are washed twice with 10-ml. portions of absolute ethanol,
and placed in a desiccator over KOH pellets for 10 hours.   Yield
is 1.95 g. (*ca.* 70%).   *Anal.* Calcd. for $[NH_4]_3[MoCl_6]$: N, 11.58;
Mo, 26.44.   Found: N, 11.5; Mo, 26.7.   The same procedure
may be followed for the rubidium and cesium compounds,
using 1.0 g. rubidium chloride or 1.5 g. cesium chloride in place of
0.5 g. ammonium chloride (see Note page 173).

Properties

The compounds are pink or red crystalline solids which are
stable indefinitely in ordinary laboratory air at 25°.   Their
chemical properties are like those of the potassium compounds,
which have previously been described.[1]

The filtrate contains an appreciable quantity of Mo^{3+}. For economy, these liquids from several preparations may be combined and evaporated to 40 ml. and the procedure repeated. The liquids should be kept in an atmosphere of nitrogen if they are to be kept longer than 1 day.

References

1. K. H. Lohmann and R. C. Young, *Inorganic Syntheses*, **4**, 97 (1953).
2. A. B. Brignole and F. A. Cotton, *ibid.*, **13**, 87 (1972).
3. L. H. Horsley, *Advan. Chem.*, **6**, 4 (1952).

34. DISODIUM HEXACHLOROPLATINATE(IV)

Submitted by LAWRENCE E. COX* and DENNIS G. PETERS*
Checked by J. HALPERN† and M. PRIBANIC†

None of the accepted preparative methods[1,2] for disodium hexachloroplatinate(IV), $Na_2[PtCl_6]$, is entirely suitable for routine use, because these procedures require reaction of a mixture of solid sodium chloride and platinum metal with chlorine gas at temperatures near 500°. However, a preparation of the desired compound is described below which yields a product of very high purity and which is much more convenient than those previously reported since only solution chemistry is involved. Unlike the corresponding ammonium and potassium salts, disodium hexachloroplatinate(IV) exhibits high solubility in water and in several nonaqueous solvents, including ethanol, methanol, and N,N-dimethylformamide—a fact which enables one to extend the range of studies of the chemistry of hexachloroplatinate(IV).

* Indiana University, Bloomington, Ind. 47401.
† The University of Chicago, Chicago, Ill. 60637.

*Procedure**

High-purity (99.99%) platinum in the form of either thin wire or foil is recommended as a starting material because of its ease of dissolution in aqua regia; all other chemicals should be of reagent-grade quality and are usable without additional purification.

A solution of hexachloroplatinic(IV) acid, H_2PtCl_6, is prepared by dissolving 4.5 g.† of platinum metal in 50–100 ml. of a 1:1 mixture of concentrated hydrochloric and nitric acids. The dissolution is carried out in a two-necked, round-bottomed, 250-ml. Pyrex flask equipped with a reflux condenser. The aqua regia solution is evaporated almost to dryness four times with successive 50-ml. portions of concentrated hydrochloric acid.

A vacuum pump, with an intervening trap, is connected to the flask, and the hydrochloric acid solution is evaporated to dryness by placing the flask in a pan of hot water. The product is a crude form of hydrated hexachloroplatinic(IV) acid; strong heating must be avoided to prevent decomposition of the solid acid. The solid acid is redissolved in a minimal volume (approximately 10 ml.) of distilled water, and disodium hexachloroplatinate(IV) is precipitated by adding a threefold excess (8.0 g.) of solid sodium chloride to the solution and cooling the mixture in an ice-water bath.‡

---

* Hexachloroplatinate(IV) may undergo appreciable photoaquation in aqueous media of low acidity and low chloride concentration. Although special steps to prevent this reaction appear to be unnecessary in the procedure described here, it is suggested that, during the course of the preparation, solutions containing hexachloroplatinate(IV) not be unduly exposed to sunlight or intense laboratory light. Accordingly, if it is convenient, as many of the operations as possible should be performed in a darkened environment.

† The checkers worked on one-third scale.

‡ The presence of excess chloride is apparently necessary for the precipitation of the sodium hexachloroplatinate(IV), since the precipitate does not form when a large excess of either sodium perchlorate or sodium nitrate is added.

The yellow-orange solid is collected on a fine-porosity sintered-glass funnel by means of suction filtration and is allowed to dry in air on the filter for a few minutes. Disodium hexachloroplatinate(IV) is separated from the excess sodium chloride by treating the solid with 50–75 ml. of hot (60°) absolute ethanol to dissolve selectively the former compound and by suction filtering the hot mixture rapidly through a fine-porosity, sintered-glass funnel. The solid disodium hexachloroplatinate(IV) is recovered by adding 20 ml. of dioxane to the ethanol filtrate and cooling the solution in an ice-water bath.*

Pure anhydrous disodium hexachloroplatinate(IV), Na_2PtCl_6, is obtained by redissolving the solid from the dioxane reprecipitation in a minimal volume of distilled water and evaporating the solution to dryness in a vacuum desiccator at room temperature and a pressure of less than 0.5 torr. To prevent loss of disodium hexachloroplatinate(IV) crystals and solution by spattering during the evaporation, the vessel containing the solution is covered with a watchglass. The anhydrous salt should be stored in a vacuum desiccator at room temperature and a pressure of less than 0.5 torr. Yield is 4.5 g. (43%).

Analysis of the compound for platinum is accomplished gravimetrically by reduction of a known weight of the anhydrous salt to metallic platinum with formic acid.[3] For the determination of chloride, 0.3 g. of the anhydrous salt is dissolved in 40 ml. of distilled water containing 250 mg. of hydrazine sulfate; the solution is boiled gently until platinum metal is formed and coagulated. Then, without removal of the platinum, the solution is made 1 F in nitric acid and is titrated potentiometrically with standard 0.2 F silver nitrate solution. *Anal.* Calcd. for Na_2PtCl_6: Pt, 42.99; Cl, 46.88. Found: Pt, 42.78; Cl, 46.73.

---

* Infrared and proton n.m.r. spectra of the recovered solid indicate that it is not pure anhydrous sodium hexachloroplatinate(IV), but that it apparently consists of a dioxane solvate; when this material is heated at 110°, it undergoes extensive decomposition.

Properties

Yellow anhydrous disodium hexachloroplatinate(IV) is quite hygroscopic and readily gains six water molecules of hydration when stored at 25° and a relative humidity of 50% or greater. Orange disodium hexachloroplatinate(IV) hexahydrate, $Na_2PtCl_6 \cdot 6H_2O$, is reconverted to the anhydrous compound by heating at 110° for one hour. The ultraviolet spectrum of hexachloroplatinate(IV) is so distinctive that it serves as a valuable criterion for the sole presence of this ion in solution as well as for the purity of disodium hexachloroplatinate(IV). In 1 F hydrochloric acid medium the spectrum exhibits a maximum at 262 nm. and a minimum at 232 nm. with molar absorptivities of 24,500 and 3080 l./mole-cm., respectively;[4] the molar absorptivity is 5090 l./mole-cm. at 300 nm.

References

1. G. Brauer (ed.), "Handbuch der Präparativen Anorganischen Chemie," p. 1368, Ferdinand Enke Verlag, Stuttgart, Germany, 1962.
2. "Gmelins Handbuch der Anorganische Chemie," Achte Auflage, No. 68, Platin, Part C, p. 148, Verlag Chemie, GmbH, Weinheim, Germany, 1953.
3. W. F. Hillebrand, G. E. F. Lundell, H. A. Bright, and J. I. Hoffman, "Applied Inorganic Analysis," 2d ed., pp. 378–379, John Wiley & Sons, Inc., New York, 1955.
4. C. K. Jørgensen, *Acta Chem. Scand.*, **10**, 518 (1956).

Chapter Seven

VARIOUS TRANSITION-METAL COMPOUNDS

35. DICHLOROTETRAKIS(2-PROPANOL)-VANADIUM(III) CHLORIDE

$$\text{VCl}_3 + 4i\text{-C}_3\text{H}_7\text{OH} \rightarrow [\text{V}(i\text{-C}_3\text{H}_7\text{OH})_4\text{Cl}_2]\text{Cl}$$

Submitted by A. T. CASEY* and R. J. H. CLARK†
Checked by T. E. BOYD,‡ W. RHINE,‡ and G. STUCKY‡

Many anhydrous metal halides will form alcoholates if the presence of water is avoided during the preparations. The reactions of alcohols with vanadium trichloride were originally thought[1] to produce hexaalcoholates, but subsequent work cast some doubt on this.[2,3] In a recent publication[4] it was established that the species formed in solution by treating vanadium trichloride with methanol, ethanol, n- and i-propyl alcohol (1- and 2-propanol), n-, i-, and s-butyl alcohol (1-butanol, 2-methyl-1-propanol, and 2-butanol), and cyclohexanol are of the type [V(ROH)$_4$Cl$_2$]Cl in each case. In some cases the species precipitated from the above solutions have the same structural type,

* University of Melbourne, Australia.
† University College, Gower St., London, W.C.1, England.
‡ University of Illinois, Urbana, Ill. 61801.

but in other cases neutral monomers of the type [VCl$_3$·3ROH] can be precipitated. The adduct of isopropyl alcohol described here is probably the easiest alcoholate to prepare, as the compound is not particularly soluble in the parent alcohol.

Procedure

The isopropyl alcohol is purified by drying over molecular sieves, refluxing with aluminium isopropoxide, and then twice fractionally distilling from aluminium isopropoxide. The unit illustrated in Fig. 8, page 166, is used for the preparation. After flaming out under vacuum, it is charged with 7.6 g. (0.048 mole) anhydrous vanadium trichloride in a nitrogen-filled dry-box. The purified isopropyl alcohol (125 ml.) is distilled onto the trichloride in a stream of dry nitrogen. The mixture is then refluxed at 85° on an oil bath, under nitrogen, for an hour and then allowed to cool slowly. Bright green crystals are deposited from the green solution. The solid is filtered on the sintered-glass disk, using a slight vacuum. The crystals may be recrystallized from fresh isopropyl alcohol, although initial slow crystallization from the original mother liquor produces a pure product. After drying under vacuum at room temperature, the adduct is stored in an evacuated manifold of the type illustrated in Fig. 9, page 166. Yield is *ca.* 15 g. or 80% based on VCl$_3$. *Anal.* Calcd. for VCl$_3$·4C$_3$H$_7$OH: C, 36.2; H, 8.1; Cl, 26.7; V, 12.8. Found: C, 36·0; H, 7.8; Cl, 26.0; V, 12.6.

Properties

Dichlorotetrakis(2-propanol)vanadium(III) chloride forms bright green crystals which are readily hydrolyzed and oxidized. Indeed, the coordinated alcohol molecules are readily displaced by most other ligands. In solution the band reported at 21,300 cm.$^{-1}$ disappears after only a few hours even in a stoppered cell, and so all spectra must be taken in a sealed cell.[4] Conductivity measurements[4] on the compound in the parent alcohol show it to be a 1:1 electrolyte in solution.

The two accessible spin-allowed ligand field bands occur at essentially the same frequency both in solution and for the solid complex (as determined from diffuse reflectance spectral measurements). Thus the solid complex is also correctly formulated as $[V(i\text{-}PrOH)_4Cl_2]Cl$. The magnetic moment of the crystalline material is 2.71 B.M. at 300°K and falls to 2.50 B.M. at 80°K, as expected, for an octahedrally coordinated vanadium(III) species with nonequivalent ligands.[4]

References

1. H. Hartmann and H. L. Schläfer, *Z. Naturforsch.*, **6a**, 754 (1951).
2. D. C. Bradley and M. L. Mehta, *Can. J. Chem.*, **40**, 1710 (1962).
3. H. Funk, G. Mohaupt, and A. Paul, *Z. Anorg. Allgem. Chem.*, **302**, 199 (1959).
4. A. T. Casey and R. J. H. Clark, *Inorg. Chem.*, **8**, 1216 (1969); "Proceedings of the 11th International Conference on Coordination Chemistry," Israel, 1968.

36. TRICHLOROBIS(TRIMETHYLAMINE)VANADIUM(III)

$$VCl_3 + 2(CH_3)_3N \xrightarrow{100°C} VCl_3 \cdot 2N(CH_3)_3$$

Submitted by A. T. CASEY,* R. J. H. CLARK,† and K. J. PIDGEON*
Checked by W. RHINE‡ and G. STUCKY‡

The recent interest in five coordination[1] has led to an intensive study of a number of transition-metal complexes which appear from their stoichiometry to contain a five-coordinate metal atom. Whereas most of this effort has been focused on the later transition elements, certain key complexes of titanium, vanadium,

* University of Melbourne, Australia.
† University College, Gower St., London, W.C. 1, England.
‡ University of Illinois, Urbana, Ill. 61801.

and chromium, for example, those of trimethylamine of the type $MX_3 \cdot 2N(CH_3)_3$ (X = Cl or Br) and of triethylphosphine of the type $MCl_3 \cdot 2P(C_2H_5)_3$ have also been studied. The following synthesis of $VCl_3 \cdot 2N(CH_3)_3$ is a modification of the one originally reported in 1957.[2]

Procedure

Anhydrous trimethylamine was purified by trap-to-trap distillation *in vacuo*. A Pyrex tube, 45 cm. long, 26 mm. o.d., and 20 mm. i.d., is flamed out under vacuum, filled with nitrogen, and transferred to an argon- or nitrogen-filled dry-box. It is then charged with 3.0 g. (0.019 mole) anhydrous vanadium trichloride and trimethylamine (9 ml., 0.15 mole). The mixture is frozen in liquid nitrogen, and the tube sealed off under vacuum. The tube must be carefully annealed. The tube and contents are then heated in a Carius furnace for 9 hours at 100°. After the tube has been cooled to room temperature it is opened under dry argon and sodium-dried benzene (70 ml.) is added. After filtration in the apparatus shown in Fig. 8, the solvent is removed under vacuum until crystallization begins (about 60 ml. is removed) and the solution set aside until precipitation is complete. The solid may be recrystallized from dry benzene if required. Yield is 3.9 g. (55% based on VCl_3). *Anal.* Calcd. for $VCl_3 \cdot 2N(CH_3)_3$: C, 26.16; H, 6.58; Cl, 38.61; N, 10.17; V, 18.48. Found: C, 27.3; H, 6.66; Cl, 37.8; N, 10.0; V, 18.9.

Properties

Trichlorobis(trimethylamine)vanadium(III) is a light mauve solid which is extremely sensitive to both moisture and oxygen. It is isostructural with $TiCl_3 \cdot 2N(CH_3)_3$ and $CrCl_3 \cdot 2N(CH_3)_3$; the latter has been shown[3] by x-ray diffraction methods to be trigonal bipyramidal with the two trimethylamine molecules lying along the trigonal axis. Other stoichiometrically similar complexes which are likewise trigonal bipyramidal include $TiBr_3 \cdot 2N(CH_3)_3$[4] and $VCl_3 \cdot 2P(C_2H_5)_3$.[5,6] Assignments for both the electronic spectra[7-9] and the V—N and V—Cl stretching frequencies[9,10] have been given. The magnetic moment of the complex is 2.69 B.M. at 291°K and is almost independent of temperature.[9] The compound sublimes at 100°.

References

1. E. L. Muetterties and R. A. Schunn, *Quart. Rev.*, **20**, 245 (1966).
2. G. W. A. Fowles and C. M. Pleass, *J. Chem. Soc.*, **1957**, 1674.
3. G. W. A. Fowles, P. T. Greene, and J. S. Wood, *Chem. Commun.*, **1967**, 971.
4. B. J. Russ and J. S. Wood, *ibid.*, **1966**, 745.
5. K. Issleib and G. Bohn, *Z. Anorg. Allgem. Chem.*, **301**, 188 (1959).
6. R. J. H. Clark and B. Sen, *Inorg. Chem.*, submitted for publication.
7. H. W. Duckworth, G. W. A. Fowles, and P. T. Greene, *J. Chem. Soc.* (A), **1967**, 1592.
8. J. S. Wood, *Inorg. Chem.*, **7**, 852 (1968).
9. A. T. Casey and R. J. H. Clark, *ibid.*, **7**, 1598–1602 (1968).
10. I. R. Beattie and T. Gilson, *J. Chem. Soc.*, **1965**, 6595.

37. VANADYL(IV) ACETATE, VO(CH₃CO₂)₂
[*Bis(acetato)oxovanadium(IV)*]

$$V_2O_5 + xs(CH_3CO)_2O \rightarrow VO(CH_3CO_2)_2$$

Submitted by RAM CHAND PAUL,* SAROJ BHATIA,* and ASHOK KUMAR*
Checked by J. T. MAGUE† and C. W. WESTON†

A method for the preparation of vanadyl(IV) acetate in non-aqueous media, as reported earlier,[1] involves solvolysis of vanadyl(V) chloride in acetic anhydride. However, a quicker

* Department of Chemistry, Punjab University, Chandigarh-14, India.
† Department of Chemistry, Tulane University, New Orleans, La. 70118.

and more convenient method for its preparation has been developed by using vanadium(V) oxide in place of vanadyl(V) chloride. The simplicity of the procedure lies in the fact that vanadium(V) oxide need not first be converted to vanadyl(V) chloride which is the starting material for the earlier method. The present method is, therefore, economical in respect to both time and cost and gives a quantitative yield of the product.

The process whereby vanadium(V) is reduced to vanadium(IV) has not been established with certainty, but it takes place spontaneously, without addition of a reducing agent.[2] The following two postulates could be made[3] to account for this:

a. $V_2O_5 + 2(CH_3CO)_2O \rightarrow 2VO(CH_3COO)_2 + \frac{1}{2}O_2$
b. $V_2O_5 + 3(CH_3CO)_2O \rightarrow 2[VO(CH_3COO)_3]$
        $2[VO(CH_3COO)_3] \rightarrow 2VO(CH_3COO)_2 + 2[CH_3COO]$
        $2[CH_3COO] \rightarrow CH_3 \cdot CH_3 + 2CO_2$

Postulate *a* may be ruled out since evolution of oxygen could not be detected while the reaction was in progress. Postulate *b* seems more likely since the evolution of carbon dioxide is indicated.

Procedure

Vanadium(V) oxide (18.2 g.) is added to acetic anhydride (50 ml.) in a round-bottomed flask fitted with a reflux condenser and a silica gel guard tube. The amount of acetic anhydride added should be in excess of that required theoretically for the reaction. This prevents the solid fluffy product from charring during refluxing of the contents. The contents are heated under reflux using an oil bath or an electric mantle maintained at $140 \pm 5°$. The start of the reaction is indicated by a change in the color from the reddish brown of vanadium(V) oxide to light gray. During the reaction, the contents may need to be shaken occasionally to prevent caking of vanadium(V) oxide. The reaction is complete in an hour as indicated by the separation of a product which does not change its gray color even on prolonged refluxing. The contents are cooled and filtered through a sintered-glass funnel. The solid product is then transferred to a 200-ml. round-bottomed flask containing carbon tetrachloride and the mixture refluxed for 10–15 minutes to remove traces of acetic anhydride from the compound. The product is again filtered, and washed with carbon tetrachloride. After it has been allowed to stand in the air for 2 hours, it is dried in vacuum at room temperature for about an hour. The excess acetic anhydride may be recovered by distillation of the filtrate. Yield is 35 g. (95%). *Anal.* Calcd. for $VO(CH_3CO_2)_2$: V, 27.57; C, 25.94; H, 3.29. Found: V, 27.3; C, 25.9; H, 3.20. Vanadium was determined by ignition to V_2O_5 and confirmed by thermogravimetric analysis.

Properties

The compound is a gray, nonhygroscopic, odorless powder. It does not melt but decomposes on heating. Its pyrolysis curve reveals that it is stable up to 214°. It loses weight between 214 and 388° and attains a constant weight at 388°, leaving a residue of vanadium(V) oxide. It is insoluble in common organic solvents, e.g., carbon tetrachloride, benzene, chloroform, and cyclohexane. It does not form any addition compounds with tertiary organic bases like pyridine, picolines, etc. Its infrared absorption spectrum (Nujol mull) has the following characteristic bands: 2845(s), 1495(s), 1450(s), 1065(w), 1040(m), 900(s), and 665(s).

References

1. R. C. Paul and A. Kumar, *J. Inorg. Nucl. Chem.*, **27**, 2537 (1965).
2. J. Vulterin, *Chem. Listy*, **53**, 392 (1959).
3. W. A. Waters, "Mechanisms of Oxidation of Organic Compounds," p. 99, Methuen & Co., Ltd., London, 1964.

38. TRIS(DIAMINE)CHROMIUM(III) SALTS

$$CrCl_3 \cdot 6H_2O + 3 \text{ diamine} \xrightarrow{Zn} [Cr(\text{diamine})_3]Cl_3 \cdot 3H_2O + 3H_2O$$

Submitted by R. D. GILLARD* and P. R. MITCHELL*
Checked by D. H. BUSCH,† C. R. SPERATI,† H. B. JONASSEN,‡
and C. W. WESTON‡

Tris(diamine) complexes of chromium(III) have been important from several viewpoints. First, the tris(ethylenediamine) complex is valuable as a synthetic intermediate,[1] the action of heat on the chloride salt giving the *cis*-dichlorobis(ethylenediamine) complex and on the thiocyanate salt giving the *trans*-bis(isothiocyanato) complex. Second, the cations are resolvable, and studies[2] of their optical activity have been fruitful in establishing relations between signs of Cotton effects and absolute stereochemistries. A large number of other studies, including kinetic and equilibrium measurements, have shown these complexes to be readily hydrolyzed to bis(ethylenediamine) complexes.[3]

Previous synthetic routes, although chemically simple, are operationally tedious:

a.[1] $Cr_2(SO_4)_3 \cdot 15H_2O \xrightarrow[\text{in vacuo}]{110^\circ \ 3 \text{ days}}$ anhydrous $Cr_2(SO_4)_3$

$$[Cr(en)_3]Cl_3 \xleftarrow[\text{HCl}]{\text{EtOH–H}_2O} [Cr(en)_3]_2(SO_4)_3 \quad \text{en} \left| \begin{array}{c} \text{under reflux} \\ \text{overnight} \end{array} \right.$$

b.[4] $CrCl_3 \cdot 6H_2O \xrightarrow{Zn–HCl} CrCl_2 \xrightarrow{\text{en}} [Cr(en)_3]Cl_2$

$$\downarrow \text{Pt–asbestos}$$

$$[Cr(en)_3]Cl_3$$

---

* University of Kent at Canterbury, Canterbury, England.
† The Ohio State University, Columbus, Ohio 43210.
‡ Tulane University, New Orleans, La. 70118.

It has been found[4] that a good yield may be obtained rapidly by allowing the commercially available green chromic chloride, $CrCl_3 \cdot 6H_2O$, in methanol to boil under reflux with ethylenediamine in the presence of metallic zinc. The product, hydrated tris(ethylenediamine)chromium(III) chloride, is obtained as a solid and is readily purified. An exactly similar procedure may be used for the complex of 1,2-propanediamine.

The zinc catalyst probably functions by generating kinetically labile chromium(II) species. The present observation is reminiscent of several others, notably (1) the ready dissolution of anhydrous chromium(III) chloride in water and other solvents only in the presence of chromium(II) ion or reducing agents, (2) the synthesis of complexes of chromium(III) by the oxidation of a preformed solution of chromium(II) ion and the appropriate ligand, e.g., $[Cr(NH_3)_6]^{3+}$,[5] $[Cr(CN)_6]^{3-}$,[6] and $[Cr(en)_3]^{3+}$,[5] and (3) the catalytic effect of charcoal on the formation of $[Cr(en)_3]^{3+}$ by the action of ethylenediamine on aqueous chromium(III) chloride.[7] It is noteworthy that in aqueous solution the use of zinc as catalyst gives only very low yields of $[Cr(en)_3]Cl_3$. The following procedure is a revised version of matter[4] extracted from the *Journal of the Chemical Society* (London) by permission.

Procedure

A single piece (*ca.* 1 g.) of granulated zinc is added to a solution of green chromium(III) chloride ($CrCl_3 \cdot 6H_2O$, 26.6 g., 0.11 mole) in methanol (50 ml.) and the mixture heated under reflux on a steam bath. Anhydrous ethylenediamine (40 ml., 36 g., 0.6 mole) is added* and the heating continued for one hour. After cooling, the solid product is collected on a Büchner or fritted-glass funnel and the piece of zinc removed. The yellow product is washed with 10% solution (50 ml.) of ethylene-

---

* This addition should be made very slowly to avoid frothing and bubbling which may cause loss of product.

diamine in methanol (until the washings are colorless) and then with ether and air-dried.

The product, tris(ethylenediamine)chromium(III) trichloride trihydrate (19.9 g., 0.051 mole, 51%), is pure; the analytical sample was further purified by recrystallization from aqueous ethanol containing a few drops of hydrochloric acid. *Anal.* Calcd. for $C_6H_{30}Cl_3CrN_6O_3$: C, 18.3; H, 7.7; Cl, 27.1; Cr, 13.3; N, 21.4%. Found: C, 18.2; H, 7.6; Cl, 27.0; Cr, 13.3; N, 21.4.

The u.v. spectrum shows: λ_{max} 457 nm., ϵ 75; λ_{max} 351 nm., ϵ 61 (literature[2c] λ_{max} 460 nm., ϵ 74; λ_{max} 353 nm., ϵ 65).

The preparation does not succeed if aqueous solutions are used, the yield then being very small. No tris(ethylenediamine)chromium(III) chloride is obtained in the absence of a piece of granulated zinc. Magnesium ribbon or granulated tin may be used instead, but the yields are lower (38 and 30%, respectively), and with granulated tin the reaction is much slower than with zinc or magnesium.

Tris(1,2-propanediamine)chromium(III) chloride dihydrate was obtained in an exactly similar manner. Yield is 20.0 g. (0.048 mole, 48%). *Anal.* Calcd. for $C_9H_{34}Cl_3CrN_6O_2$: C, 25.9; H, 8.1; Cl, 25.6; Cr, 12.5; N, 20.2%. Found: C, 25.7; H, 8.1; Cl, 25.8; Cr, 12.6; N, 20.2.

The absorption spectrum showed: λ_{max} 459 nm., ϵ 76; λ_{max} 352 nm., ϵ 63 (literature[2c] λ_{max} 460 nm., ϵ 71; λ_{max} 353 nm., ϵ 56)

References

1. C. L. Rollinson and J. C. Bailar, *Inorganic Syntheses*, **2**, 196 (1946); cf. *ibid.* **13**, 232 (1972).
2. (a) J. P. Mathieu, *J. Chim. Phys.*, **33**, 78 (1936); (b) J. H. Dunlop and R. D. Gillard, *J. Inorg. Nucl. Chem.*, **27**, 361 (1965); (c) A. J. McCaffery, S. F. Mason, and R. E. Ballard, *J. Chem. Soc.*, **1965**, 2883.
3. H. L. Schläfer and O. Kling, *Z. Anorg. Allgem. Chem.*, **287**, 296 (1956).
4. R. D. Gillard and P. R. Mitchell, *J. Chem. Soc.* (A), **1968**, 2129.
5. D. Berman, G. Bokerman, and R. W. Parry, *Inorganic Syntheses*, **10**, 35 (1967).
6. H. Moissan, *Ann. Chim. Phys.*, **25**, 401 (1882).
7. J. C. Bailar and J. B. Work, *J. Am. Chem. Soc.*, **67**, 176 (1945).

39. BIS [cis-1,2-DICYANOETHENE-1,2-DITHIOLATO-(1− OR 2−)] COMPLEXES OF COBALT AND IRON

Submitted by J. BRAY,* J. LOCKE,* J. A. McCLEVERTY,* and D. COUCOUVANIS†
Checked by J. P. FACKLER‡ and R. JACKSON‡

There has been considerable interest in the chemistry and electronic structures of cobalt and iron complexes of *cis*-1,2-disubstituted ethene-1,2-dithiol[1,2] and their Lewis base adducts.[3−5] The complexes, and their Lewis base adducts which may contain pyridine, phosphines, NO, dipyridyl (2,2′-bipyridine), etc., are capable of undergoing reversible one-electron transfer reactions, thereby generating a series of complex ions differing from each other by only one unit of electric charge.

Some of the most readily accessible groups of cobalt and iron complexes containing dithiolato ligands are those obtained from *cis*-1,2-dicyanoethene-1,2-dithiolate, $[S_2C_2(CN)_2]^{2-}$, and the syntheses of $[(n\text{-}C_4H_9)_4N]_2[CoS_4C_4(CN)_4]$ and $[(n\text{-}C_4H_9)_4N]_2[CoS_4C_4(CN)_4]_2$ have been described, although not in detail, in a previous volume.[6] The latter complex was obtained from the former by its oxidation using $[NiS_4C_4(CF_3)_4]$, but a more economical and faster preparation for salts of $[CoS_4C_4(CN)_4]_2{}^{2-}$ has been devised. Also a convenient modification for the procedure used to obtain salts of $[CoS_4C_4(CN)_4]^{2-}$ is presented. The synthesis of salts of $[FeS_4C_4(CN)_4]_2{}^{2-}$, from which many Lewis base adducts can be prepared, is described in full.

A. SODIUM CYANODITHIOFORMATE-3N,N-DIMETHYLFORMAMIDE

$$NaCN + CS_2 + HCON(CH_3)_2 \rightarrow NaS_2CCN \cdot 3HCON(CH_3)_2$$

One of the key factors in the speedy syntheses of *cis*-dicyano-ethylene-1,2-dithiolato complexes is the rapid preparation of the

---

* Department of Chemistry, The University, Sheffield, England.

† Department of Chemistry, Case Western Reserve University, Cleveland, Ohio 44106. Present address: The University of Iowa, Iowa City, Iowa 52240.

‡ Chemistry Department, Case Western Reserve University, Cleveland, Ohio 44106.

dithiolato ligand precursor, sodium cyanodithioformate (NaS$_2$-CCN). The synthesis of this material has been described elsewhere,[6,7] but an accelerated and improved procedure has been devised.

Procedure

Carbon disulfide (7.6 g.) is added dropwise, over a period of about 10 minutes, to a solution of 4.9 g. sodium cyanide in 50 ml. N,N-dimethylformamide in a 500-ml. conical flask. During the addition the mixture is maintained at about 0° by shaking in an ice bath. After the addition is complete, the dark brown reaction mixture is treated with 50 ml. n-butanol and warmed, with (magnetic) stirring, to about 100° for 20 minutes. The mixture is then filtered while hot, treated with an additional 50 ml. n-butanol, and cooled in ice. Approximately 30 g. of the crystalline NaS$_2$CCN·3HCON(CH$_3$)$_2$ is obtained. This material decomposes in a few days unless maintained in scrupulously dry conditions and, in any case, should be used as soon as possible after preparation.

B. THE FORMATION OF THE LIGAND SOLUTION

$$2S_2CCN^- \rightarrow S_2C_2(CN)_2{}^{2-} + 2S$$

Sodium cyanodithioformate readily, and quickly, dimerizes in water with elimination of sulfur in the presence of transition-metal ions, forming directly, in this process, a dithiolato-metal complex. The general procedures described below have been used in the preparation of most of the bis(dicyanoethenedithiolato) complexes described previously[6] and also of some tris(cis-1,2-dicyanoethene-1,2-dithiolato) complexes of the first-row transition metals. Although the yields of these complexes obtained by this method are somewhat lower than those described previously, the method possesses the advantage of speed and relative convenience.

Procedure

The weighed sample of sodium cyanodithioformate is dissolved in 50 ml. water and heated on a steam bath for 15–20 minutes. The partly cloudy yellow-brown solution is then filtered through Kieselguhr or Celite to remove precipitated sulfur, and the residue is washed with 20 ml. water, the washings being added to the filtrate. This filtrate contains some $Na_2S_2C_2(CN)_2$, undimerized NaS_2CCN and *N,N*-dimethylformamide, and is referred to as the *ligand solution;* the amount of cyanodithioformate used to make up this ligand solution appears in parentheses after "ligand solution."

Note

The *salt* $Na_2S_2C_2(CN)_2$, may be obtained in reasonable yields (70%) from $NaS_2CCN \cdot 3HCON(CH_3)_2$ by refluxing the latter in chloroform on a steam bath for 8 hours.[8] The product contains a small amount of contaminatory sulfur but is quite suitable for most synthetic purposes.

C. BIS(TETRAPHENYLPHOSPHONIUM) BIS[*cis*-1,2-DICYANOETHENE-1,2-DITHIOLATO(2−)]COBALTATE(2−)

$$Co^{2+} + 2[S_2C_2(CN)_2]^{2-} + 2[(C_6H_5)_4P]^+ \xrightarrow{SO_3^{2-}} [(C_6H_5)_4P]_2[CoS_4C_4(CN)_4]$$

Procedure

Two and one-half grams of cobalt acetate tetrahydrate in 40 ml. water is treated with the ligand solution (13.8 g.). To the resulting red-brown solution 5.0 g. sodium sulfite in 30 ml. warm water is added, and the mixture is heated on a steam bath for 20 minutes. After this time, the mixture is filtered and the filtrate cooled, and to it is added 8.0 g. tetraphenylphosphonium bromide dissolved in 30 ml. warm ethanol. The black tarry precipitate which forms is collected by filtration, washed with water, and extracted into hot *N,N*-dimethylformamide, preferably under a stream of nitrogen. Crystallization of the complex is effected by dropwise addition of diethyl ether to a hot solution of the compound in aqueous acetone containing sodium sulfite. On standing for several hours, preferably under nitrogen, fine black crystals separate and are collected by filtration, washed with ethanol and ether, and air-dried, m.p. 247–249° (uncorrected, with decomp.). The yield is between 6.5 and 6.7 g. [59–66% based on $Co(OCOCH_3)_2 \cdot 4H_2O$].

Notes

1. The complex salt $[(C_2H_5)_4N]_2[CoS_4C_4(CN)_4]$ may be obtained similarly, the black precipitate being extracted into hot acetone containing aqueous sodium sulfite, and recrystallization may be effected from this acetone extract by addition of water containing sodium sulfite or by the procedure described above. If sulfite is omitted during these preparations, gradual formation of $[CoS_4C_4(CN)_4]^-$ or $[CoS_4C_4(CN)_4]_2{}^{2-}$ occurs. The complex, m.p. 197–200° (uncorrected, with decomp.), is obtained in 61% yields.

2. Tetraphenylphosphonium bromide is obtained in 45% yields from the reaction between 100 g. pulverized aluminum trichloride, 105 g. powdered triphenylphosphine, 60 g. pure bromobenzene, and 70 g. potassium bromide using the method outlined by Chatt and Mann.[9]

3. A white solid which may contaminate the desired product during the recrystallization stages may be removed by systematic washing with water.

4. The aqueous acetone solutions containing sodium sulfite are simply prepared by dissolving the latter in acetone containing the minimum amount of water necessary to effect its dissolution.

Properties

The complex is readily soluble in polar organic solvents affording a red or red-brown solution. On prolonged exposure to air, however, the solutions gradually adopt a green-brown hue due to the formation of $[CoS_4C_4(CN)_4]^-$ or $[CoS_4C_4(CN)_4]_2{}^{2-}$. A single-crystal x-ray analysis of $[(n\text{-}C_4H_9)_4N]_2[CoS_4C_4(CN)_4]$ has established[10] the planarity of the anion (idealized D_{2h} symmetry). The complex, which has been isolated with many different cations, is paramagnetic, having magnetic moments in the range 1.87–2.22 B.M., consistent with one unpaired electron.[11]

The complex reacts with nitric oxide in solution affording the orange-red mononitrosyl $[Co(NO)S_4C_4(CN)_4]^{2-}$, which is diamagnetic, and has been obtained as the salt of several different cations.[3]

D. BIS(TETRAPHENYLPHOSPHONIUM) TETRAKIS(*cis*-1,2-DICYANOETHENE-1,2-DITHIOLATO)DICOBALTATE(2−)

$$2[(C_6H_5)_4P]_2[CoS_4C_4(CN)_4] + I_2 \rightarrow$$
$$[(C_6H_5)_4P]_2[CoS_4C_4(CN)_4]_2 + 2[(C_6H_5)_4P]I$$

Procedure

Two grams of bis(tetraphenylphosphonium) bis(*cis*-1,2-dicyanoethene-1,2-dithiolato)cobaltate(2−) is dissolved in 20 ml. N,N-dimethylformamide and treated with 0.5 g. iodine dissolved in 10 ml. acetone. The solution, which is initially colored deep red-brown, immediately darkens, becoming green-brown, and on dropwise addition of water (usually about 15 ml.), becomes cloudy and a black solid precipitates gradually. The complex is crystallized by dissolution in 50 ml. acetone containing 10 ml. N,N-dimethylformamide followed by precipitation using diethyl ether. Fine black crystals separate and are collected by filtration, washed with ethanol and ether, and air-dried; m.p. 270–273° (uncorrected, with decomp.). The yield of the complex is 1.3 g. {95% based on $[(C_6H_5)_4P]_2[CoS_4C_4(CN)_4]$}.

Properties

The complex is readily soluble in strongly polar solvents giving a green-brown or green solution. Voltammetric examination[2] of dichloromethane solutions containing this anion reveals that it is a member of the electron-transfer series:

$$[CoS_4C_4(CN)_4]^{2-} \rightleftarrows [CoS_4C_4(CN)_4]_2{}^{2-} \rightleftarrows [CoS_4C_4(CN)_4]_2{}^{-}$$

The dimeric nature of the complex has not been proved conclusively in the solid state, but conductometric experiments[2,12] have shown that it is largely dimeric in acetonitrile solution. Reduction of the complex to the monomeric dianion is conveniently achieved using sodium sulfite in aqueous solution. The compound is diamagnetic in the solid state and in all solutions except dimethyl sulfoxide where it has a magnetic moment of 2.8 B.M.[13]

Treatment of the complex with donor molecules such as pyridines, phosphines, bipyridyl (2,2′-bipyridine), etc., results in the cleavage of the dimer and formation of diamagnetic five- and six-coordinate adducts.[4] Nitric oxide reacts with the complex forming the unstable green mononitrosyl $[Co(NO)S_4C_4(CN)_4]^-$ which has one unpaired electron.[3] Treatment with $Na_2S_2C_2(CN)_2$ results in the formation of $[CoS_6C_6(CN_6)]^{3-}$ which can be isolated as the tetraphenylphosphonium salt.[14]

E. BIS(TETRAETHYLAMMONIUM) TETRAKIS(cis-1,2-DICYANOETHENE-1,2-DITHIOLATO)DIFERRATE(2—)

$$2Fe^{3+} + 4[S_2C_2(CN)_2]^{2-} + 2[(C_2H_5)_4N]^+ \rightarrow$$
$$[(C_2H_5)_4N]_2[FeS_4C_4(CN)_4]_2$$

Procedure

To 2.7 g. iron trichloride hexahydrate in 50 ml. water is added the ligand solution (13.8 g.). After 10 minutes, the black precipitate which forms is collected by filtration with the aid of Kieselguhr or Celite, washed thoroughly with 200 ml. water, and extracted into 50 ml. acetone. The acetone extract is filtered, the residue washed with 10 ml. acetone, and the washings added to the filtrate. To the deep red extract is added 2.1 g. tetraethylammonium bromide in 20 ml. ethanol, and the mixture is evaporated on a steam bath almost to dryness. On cooling and standing for several hours at 0°, large black crystals separate which are collected by filtration, washed with ether, and air-dried. The yield of the product, which is sufficiently pure for most practical purposes is 3.1 g. (64% based on $FeCl_3 \cdot 6H_2O$). Recrystallization may be effected by extraction of the black crystals into hot acetone, followed by addition of an equal volume of isopropyl alcohol or isobutyl alcohol and gradual evaporation of this mixture on a steam bath until crystallization begins. On cooling and standing, fine black crystals are formed which may be collected by filtration, washed with isopropyl alcohol or isobutyl alcohol and ether, and air-dried; m.p. 290° (uncorrected, with decomp.). After recrystallization the yield is 2.2–2.5 g. (47–53% based on $FeCl_3 \cdot 6H_2O$).

Notes

1. The black solid isolated initially is probably $Na_2[FeS_4C_4(CN)_4]_2$, which may be used for preparing other heavy organic cation salts of this anion or in the formation of $[Fe(NO)S_4C_4(CN)_4]^-$.[3]

2. The complex $[(n\text{-}C_4H_9)_4N]_2[FeS_4C_4(CN)_4]_2$ has been prepared by the method described above, the yield before recrystallization being 71%. Further, $[(C_6H_5)_4P]_2[FeS_4C_4(CN)_4]_2$ may be obtained in low yields by the same procedure, providing that the stoichiometries are rigidly adhered to and air is rigorously excluded from the reaction mixture.[14]

Properties

The complex is readily soluble in strongly polar solvents and in chloroform and dichloromethane giving red-brown or red solutions. Voltammetric studies[2] have shown that this dianion is a member of an electron-transfer series similar to its cobalt analog:

$$[\text{FeS}_4\text{C}_4(\text{CN})_4]^{2-} \rightleftarrows [\text{FeS}_4\text{C}_4(\text{CN})_4]_2^{2-} \rightleftarrows [\text{FeS}_4\text{C}_4(\text{CN})_4]_2^{-}$$

of which only the central member can be prepared easily. The structure of the complex in the solid state, as the $[(n\text{-}\text{C}_4\text{H}_9)_4\text{N}]^+$ salt, has been determined by x-ray techniques,[15] which show conclusively that the anion is dimeric, consisting of two nearly planar $\text{FeS}_4\text{C}_4(\text{CN})_4$ units linked by iron–sulfur bonds which effectively render each iron atom five-coordinate with respect to sulfur, the coordination geometry around the metal atoms being nearly square-pyramidal. The nature of the complex ion in solution is not so well understood but there is some evidence for dissociation into monomers.[2,4] In the solid state the complex has a magnetic moment at room temperature consistent with one unpaired spin per iron atom[2,16] but in solution, particularly in pyridine, dimethyl sulfoxide, and N,N-dimethylformamide the moment is consistent with three unpaired electrons.[17]

Treatment of the complex with donor molecules such as pyridines, phosphines, bipyridyl (2,2′-bipyridine), etc., results in the formation[4] of paramagnetic five- and six-coordinate adducts. With nitric oxide the dianion is cleaved, forming the diamagnetic mononitrosyl $[\text{Fe(NO)S}_4\text{C}_4(\text{CN})_4]^-$, and with $[\text{S}_2\text{C}_2(\text{CN})_2]^{2-}$ in air $[\text{FeS}_6\text{C}_6(\text{CN})_6]^{2-}$ is formed.[14]

References

1. J. A. McCleverty, *Prog. Inorg. Chem.*, **10**, 50 (1968); G. N. Schrauzer, *Transition Metal Chemistry*, **4**, 299 (1968); H. B. Gray, *ibid.*, **1**, 240 (1965).
2. A. L. Balch, I. G. Dance, and R. H. Holm, *J. Am. Chem. Soc.*, **90**, 1139 (1968).

3. J. A. McCleverty, N. M. Atherton, J. Locke, E. J. Wharton, and C. J. Winscom, *ibid.*, **89**, 6082 (1967).
4. N. G. Connelly, J. A. McCleverty, and C. J. Winscom, *Nature*, **216**, 999 (1967); J. A. McCleverty, M. Atherton, N. G. Connelly, and C. J. Winscom, *J. Chem. Soc.* (A), **1969**, 2242.
5. G. N. Schrauzer, V. P. Mayweg, H. W. Finck, and W. Heinrich, *J. Am. Chem. Soc.*, **88**, 4604 (1966); A. L. Balch, *Inorg. Chem.*, **6**, 2158 (1967).
6. A. Davison and R. H. Holm, *Inorganic Syntheses*, **10**, 8 (1967).
7. G. Bahr and G. Schleitzer, *Chem. Ber.*, **88**, 1771 (1955).
8. J. Locke and J. A. McCleverty, *Inorg. Chem.*, **5**, 1157 (1966).
9. J. Chatt and F. G. Mann, *J. Chem. Soc.*, **1940**, 1192.
10. J. D. Forrester, A. Zalkin, and D. H. Templeton, *Inorg. Chem.*, **3**, 1500 (1964).
11. A. Davison, N. Edelstein, R. H. Holm, and A. H. Maki, *J. Am. Chem. Soc.*, **85**, 3049 (1963).
12. A. Davison, D. V. Howe, and E. T. Shawl, *Inorg. Chem.*, **6**, 458 (1967).
13. C. H. Langford, E. Billig, S. I. Shupack, and H. B. Gray, *J. Am. Chem. Soc.*, **86**, 2958 (1964).
14. J. A. McCleverty, J. Locke, E. J. Wharton, and M. Gerloch, *J. Chem. Soc.* (A), **1968**, 816.
15. W. C. Hamilton and I. Bernal, *Inorg. Chem.*, **6**, 2003 (1967).
16. J. Weiher, L. R. Melby, and R. E. Benson, *J. Am. Chem. Soc.*, **86**, 4339 (1964).
17. R. Williams, J. H. Waters, and H. B. Gray, *ibid.*, **88**, 43 (1966).

40. RESOLUTION OF THE (ETHYLENEDIAMINE)-BIS(OXALATO)COBALTATE(III) ION

Submitted by J. H. WORRELL*
Checked by E. B. KIPP† and R. A. HAINES†

Because the optically active forms of the (ethylenediamine)bis-(oxalato)cobaltate(III) ion have been, and will continue to be, excellent resolving agents for many cationic cobalt(III) complexes, it is only proper that a detailed resolution procedure be developed to produce both enantiomorphic forms in useful quantities.

Dwyer, Reid, and Garvan have reported a general method for the resolution of $[Co(en)(ox)_2]^-$. In their resolution procedure,[1] they fail to indicate such data as volumes, yields, and

* University of South Florida, Tampa, Fla. 33620.
† The University of Western Ontario, London, Canada.

the details of their fractional crystallization schemes. Although their directions allow one intuitively skilled in the art to obtain an undisclosed quantity of one optically active antipode, the method used to obtain the other optically active form is complicated by the incorrect use of d and l rotational designations. The preparation of materials used in the resolution as well as the resolution itself is a formidable time-consuming task. This author feels the following experimental detail, with a view toward clarification, will enable future investigators to conserve time and energy in the resolution of the (ethylenediamine)bis-(oxalato)cobaltate(III) ion.

Reagent Preparations

Racemic sodium and calcium (ethylenediamine)bis(oxalato)-cobaltate(III) are prepared by the method given by Dwyer, Reid, and Garvan.[1]

trans-Dichlorobis(ethylenediamine)cobalt(III) chloride is prepared by the method given by Bailar.[2]

cis-Bis(ethylenediamine)dinitrocobalt(III) nitrite is prepared by dissolving *trans*-$[Co(en)_2Cl_2]Cl$ (20 g.) in 50 ml. of water which is then warmed to 40–50° with stirring. To this solution is added potassium nitrite (22 g.) dissolved in 50 ml. of water. The solution is filtered and placed in an ice bath. Yellow *cis*-$[Co(en)_2(NO_2)_2]NO_2$ precipitates as the side of the beaker is scratched with a glass rod. The product (12 g.) is collected by filtration, washed with ethanol and ether, and air dried.

An alternate method has been developed by Harbulak and Albinak.[3]

The *cis*-$[Co(en)_2(NO_2)_2]NO_2$ is resolved by the method developed by Dwyer and Garvan[4] using a fourfold increase in reagent quantities.

Procedure

Silver oxalate (3.5 g.) is prepared from an aqueous solution of oxalic acid and silver nitrite and collected by filtration. The solid silver oxalate is added to $(+)_{546}$-$[Co(en)_2(NO_2)_2]Br$ (8.25 g., 1.9 × 10^{-2} mole, $[\alpha]_{546}$ = +110 ± 10°) in 300 ml. of water at 55°, shaken vigorously for 3 minutes, and filtered. Racemic $Ca[Co(en)(ox)_2]_2$ (7.05 g., 1.7 × 10^{-2} mole) is added to the orange-yellow filtrate, and the solution is stirred mechanically for 30 minutes at 50–55°. The calcium oxalate which forms is quickly filtered off, and solid racemic $Na[Co(en)(ox)_2]$ (6.93 g., 2.0 × 10^{-2} mole) is added to the red-brown filtrate. The mixture is rapidly stirred, and the temperature is maintained at 50–55° for 20 minutes. The red-brown diastereoisomer of $(+)_{546}$-$[Co(en)_2(NO_2)_2](+)_{546}$-$[Co(en)(ox)_2]$ (5.0 g.) is collected by rapid filtration at 50°, washed with ethanol, and dried *in vacuo* at 25°. An aqueous solution containing 21.4 mg./100 ml. gives α = +0.212 whence $[\alpha]_{546}$ = +991 ± 15°. *Anal.* Calcd. for $(+)_{546}$-$[Co(C_4H_{16}N_4)(NO_2)_2](+)_{546}$-$[Co(C_2H_8N_2)(C_4O_8)]$: C, 21.21; H, 4.24; N, 19.79. Found: C, 21.33; H, 4.31; N, 19.90.

The filtrate is transferred to a graduated 500-ml. beaker and treated as indicated in the flowsheet. By cooling the filtrate to 35°, a second fraction of red-brown solid (0.55 g, $[\alpha]_{546}$ = −292 ± 10°) is collected. The filtrate is stirred, heated to 45°, and the solvent evaporated under a stream of air until the volume is 200 ml. The volume is measured with the stirring apparatus turned off. The solution is filtered giving 0.35 g. of solid with $[\alpha]_{546}$ = −875 ± 12°. Again the volume is reduced at 45° under an airstream to 125 ml. The mixture is filtered yielding 3.75 g. of red-brown solid with $[\alpha]_{546}$ = −880 ± 16°. The third and fourth fractions contain $(+)_{546}$-$[Co(en)_2(NO_2)_2]$-$(-)_{546}$-$[Co(en)(ox)_2]$. A portion of the diastereoisomer is recrystallized from hot ethanol–water solution, collected by filtration, washed with ethanol and ether, and dried *in vacuo* at 25°. An aqueous solution containing 26.86 mg./100 ml. gives α = −0.237 whence $[\alpha]_{546}$ = −890 ± 15°.

Additional fractions of diastereoisomer and starting material are obtained as the volume is slowly reduced to 35 ml.; however, the rotations are not sufficient to warrant recovery of optically pure $[Co(en)(ox)_2]^-$. The resolving agent present in these latter fractions is recovered as the iodide by grinding the solid with sodium iodide in a water slurry followed by filtration. Ethanol is added to the purple-red filtrate and $Na[Co(en)(ox)_2]$ precipitates. It can be purified by recrystallization from water–ethanol. *Anal.* Calcd. for $(+)_{546}$-$[Co(C_4H_{16}N_4)(NO_2)_2](-)_{546}$-$[Co(C_2H_8N_2)(C_4O_8)]$: C, 21.21; H, 4.24; N, 19.79. Found: C, 21.15; H, 4.14; N, 19.99.

The diastereoisomers are broken up and the resolved complex recovered as the sodium salt as indicated below. The first fraction (5.0 g.) of $(+)_{546}$-$[Co(en)_2(NO_2)_2](+)_{546}$-$[Co(en)(ox)_2]$ is ground with 30 ml. of warm (35–40°) water containing 6.0 g. of sodium iodide. The solution is cooled and the insoluble yellow $(+)_{546}$-$[Co(en)_2(NO_2)_2]I$ collected by filtration. Absolute ethanol (15 ml.) is added to the purple-red filtrate, and the solution is cooled in the refrigerator for 6 hours. The first crop of red-violet crystals (2.0 g.) is collected by filtration, washed with acetone, and air-dried. More absolute ethanol is added to the filtrate, and cooling is continued for 4 hours. Three additional fractions are collected in this way totalling 0.85 g. of complex. An aqueous solution of the first fraction containing 32.7 mg./100 ml. gave $\alpha = +0.471$ whence $[\alpha]_{546} = +1440 \pm 10°$ ($+1350°$ by checkers). The rotation of the 1-hydrate given by Dwyer et al. is $[\alpha]_{546} = +1400$.[1] The $(+)_{546}$-$Na[Co(en)-(ox)_2]$ obtained above was not recrystallized prior to analysis. *Anal.* Calcd. for $(+)_{546}$-$Na[Co(C_2H_8N_2)(C_4O_8)]\cdot3\frac{1}{2}H_2O$: C, 18.90; H, 4.05; N, 7.35. Found: C, 19.10; H, 4.17; N, 7.18.

The other antipode, $(-)_{546}$-$Na[Co(en)(ox)_2]$, is isolated as follows. Fractions three and four are combined giving 4.0 g. of $(+)_{546}$-$[Co(en)_2(NO_2)]\cdot(-)_{546}$-$[Co(en)(ox)_2]$ which is ground up with 5.0 g. of sodium iodide dissolved in 30 ml. of water at 40°. The mixture is filtered to remove yellow $(+)_{546}$-$[Co(en)_2-(NO_2)_2]I$ (2.9 g.), and the filtrate is treated as before with absolute ethanol to cause $(-)_{546}$-$Na[Co(en)(ox)_2]$ to crystallize. The ethanol is added slowly with stirring until the very first signs of crystallization (clouding) occur; then the solution is cooled for 4 hours. If crystallization is not evident, more ethanol is carefully added and the cooling continued. Two fractions of red-violet crystals are obtained, totalling 2.60 g. These are washed with acetone and air-dried. An aqueous solution containing 26.7 mg./100 ml. gives $\alpha = -0.385$ whence $[\alpha]_{546} = -1442 \pm 16°$ ($-1330°$ by checkers). The rotation reported by Dwyer et al. for $(-)_{546}$-Na[Co(en)(ox)$_2$]·H$_2$O is $-1400°$.[1] This sample was not recrystallized prior to analysis. *Anal.* Calcd. for $(-)_{546}$-Na[Co(C$_2$H$_8$N$_2$)(C$_4$O$_8$)]·3$\frac{1}{2}$H$_2$O: C, 18.90; H, 4.05; N, 7.35. Found: C, 19.16; H, 3.86; N. 7.30.

The complete resolution is possible using identical reagent quantities and $(-)_{546}$-[Co(en)$_2$(NO$_2$)$_2$]$^+$, having $[\alpha]_{546} = -115 \pm 10°$ as the resolving agent. The results are indicated in the flowsheet for the isolation of the two diastereoisomers. As in the first procedure, at least 2.0 g. of each optically pure enantiomorph is obtained.* Figure 10 gives the partial fractional crystallization schemes corresponding to the isolation of diastereoisomers when $(+)_{546}$-[Co(en)$_2$(NO$_2$)$_2$]$^+$ and $(-)_{546}$-[Co(en)$_2$(NO$_2$)$_2$]$^+$ are utilized as resolving agents, respectively. By carefully controlling the temperature and volume of the mixture, one can obtain a clean separation of $(+)_{546}$-[Co(en)$_2$(NO$_2$)$_2$]-$(+)_{546}$-[Co(en)(ox)$_2$] from the more soluble diastereoisomer, $(+)_{546}$-[Co(en)$_2$(NO$_2$)$_2$]$(-)_{546}$-[Co(en)(ox)$_2$]. A similar situation exists when $(-)_{546}$-[Co(en)$_2$(NO$_2$)$_2$]$^+$ is employed as the resolving agent. Optically pure Na[Co(en)(ox)$_2$] is recovered from the filtrate as shown in Fig. 11. Both optically pure isomers were found to be 3$\frac{1}{2}$-hydrates having equal and opposite specific rotations at 546 nm. of $(+)$ and $(-)$ 1440 $\pm$ 16°, respectively.

Properties

The optical isomers of Na[Co(en)(C$_2$O$_4$)$_2$] are stable indefinitely in the solid state, when stored in the dark. Photo-

---

* The checkers obtained approximately 1.2 g. of pure product.

$[Co(en) (ox)_2]^- + (+)_{546} [Co(en)_2 (NO_2)_2]^+$

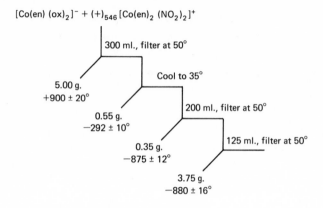

$[Co(en) (ox)_2]^- + (-)_{546} [Co(en)_2 (NO_2)_2]^+$

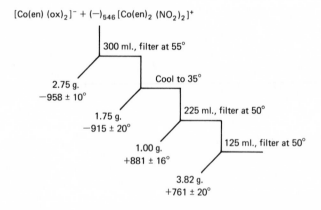

All rotations are at the
mercury green line, 546 mμ.

Fig. 10. Crystallization scheme for diastereoisomer isolation.

racemization, especially in direct sunlight, occurs in both the solid state and in aqueous solution. Slow racemization accompanied by some decomposition occurs in boiling aqueous solution in the dark and in cold 0.1 N sodium hydroxide solution.[1]

Optical rotatory dispersion and circular dichroism spectra

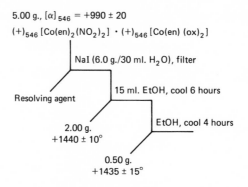

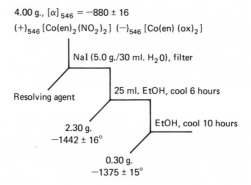

All rotations are at the
mercury green line, 546 mμ.

**Fig.   11.   Recovery   of   optically   pure
Na[Co(en)(ox)₂].**

have been reported in the literature.[5]   The salt Na[Co(en)-
$(C_2O_4)_2]\cdot H_2O$ exhibits two absorption bands in the visible region
at 541 and 384 nm. with molar extinction coefficients of 112
and 177, respectively.

References

1. F. P. Dwyer, I. Reid, and F. Garvan, *J. Am. Chem. Soc.*, **83**, 1285 (1961).
It should be noted that the yield of racemic $Na[Co(C_2O_4)_2en]$ obtained from

the method described in reference 1 is significantly lower than quoted (by *ca.* 50%).

2. J. C. Bailar, Jr., *Inorganic Syntheses*, **2**, 222 (1946); cf. *ibid.* **13**, 232 (1972).
3. F. P. Harbulak and M. J. Albinak, *J. Inorg. Nucl. Chem.*, **25**, 232 (1963).
4. F. P. Dwyer and F. L. Garvan, *Inorganic Syntheses*, **6**, 195 (1960).
5. B. E. Douglas, R. A. Haines, and J. G. Brushmiller, *Inorg. Chem.*, **2**, 1194 (1963).

41. NITRITO COMPLEXES OF NICKEL(II) AND COBALT(II)

Submitted by RICHARD H. BUCHI,* LAILA EL-SAYED,†
and RONALD O. RAGSDALE*
Checked by F. BASOLO,‡ D. DIEMENTE,‡ and D. V. STYNES‡

The nitrite ion NO_2^- can coordinate to metal atoms in four distinct ways, giving rise to the following "ligand isomers": (1) by formation of a metal–nitrogen bond, giving nitro complexes; (2) by formation of a metal–oxygen bond, giving nitrito complexes; (3) by acting as a bridging group bridging through a nitrogen atom and an oxygen atom; and (4) by functioning as a chelating group by bonding through both oxygen atoms to a metal ion.   Nitrito complexes of Co(III) were first prepared by Jorgensen[1] who noted their rearrangement to the corresponding nitro complexes.   The rates of nitrito–nitro isomerization of the $M(NH_3)_5ONO^{n+}$ complexes, where $M = $ Co(III), Rh(III), Ir(III), and Pt(IV), have been studied.[2,3]   Palladium(II) and cobalt(III) have been shown to form NO_2 bridges.[4]   Goodgame[5] first prepared chelating NO_2 groups in some cobalt(II) complexes.   Complexes of the type $CoL_2(NO_2)_2$, where L represents

* University of Utah, Salt Lake City, Utah 84112.
† University of Alexandria, Alexandria, Egypt.
‡ Northwestern University, Evanston, Ill. 60201.

a substituted pyridine *N*-oxide or a substituted quinoline *N*-oxide, have been characterized.[6] Some nitrito complexes of nickel(II) with disubstituted ethylenediamines and pyridine have been isolated.[7] These compounds are quite stable with respect to isomerization to the nitro compounds in the solid state. Stable nitrito complexes of nickel(II) also have been prepared with some substituted aminomethylpyridine and piperidine bases.[8]

The syntheses presented here are examples of two types of "ligand isomers": (1) Nitrito complexes, [Ni{2-[(methylamino)-methyl]pyridine}$_2$(ONO)$_2$], and [Ni{2-[(methylamino)methyl]-piperidine}$_2$(ONO)$_2$]; (2) Chelating nitrite groups, [Co(2,4,6-trimethylpyridine *N*-oxide)$_2$(NO$_2$)$_2$], [Co(2,6-dimethylpyridine *N*-oxide)$_2$(NO$_2$)$_2$], and [Co(4-methylquinoline *N*-oxide)$_2$(NO$_2$)$_2$].

Preparation of Nickel(II) Nitrite Solution

Twenty-nine grams (0.1 mole) Ni(NO$_3$)$_2$·6H$_2$O is dissolved in a minimum amount of methanol (50 ml.). Sodium nitrite (13.8 g., 0.2 mole) is added slowly with stirring to the green nickel nitrate solution and placed in a Dry Ice–acetone bath. The resulting sodium nitrate precipitate is filtered in a large Büchner funnel. The dark green filtrate is diluted with methanol to 100 ml. in a volumetric flask. The resulting solution is a one molar solution of nickel nitrite.

Preparation of Nitrito Complexes

1. [Ni{2-[(methylamino)methyl] pyridine}$_2$(ONO)$_2$]

Four and eighty-eight hundredths grams (0.04 mole) of 2-[(methylamino)methyl]pyridine (obtained from Aldrich Chemical) is added to 20 ml. (0.02 mole) of nickel nitrite. A dark purplish oil is obtained after removal of the methanol by rotary evaporation. The oil is dissolved in a minimum amount of absolute ethanol. The solution is extracted twice with ether. Upon addition of more ether, blue-violet crystals are obtained by trituration. The purified compound melts at 206°. Yield is 3.91 g. (52%). *Anal.* Calcd. for $C_{14}H_{20}N_6NiO_4$: C, 42.56; H, 5.10; N, 21.27. Found: C, 42.74; H, 5.36; N, 20.89.

2. [Ni{2-[(methylamino)methyl]piperidine}₂(ONO)₂]

Add 2.54 g. (0.02 mole) of 2-[(methylamino)methyl]piperidine (obtained by reduction of 2-[(methylamino)methyl]pyridine with sodium and alcohol*) to 10 ml. (0.01 mole) of nickel nitrite. The bluish-purple solution is concentrated and left overnight. The resulting blue crystals are recrystallized in absolute ethanol, filtered, washed with cold ethanol and dry ether, and dried *in vacuo*. The compound melts at 223–225°. Yield is 2.43 g. (59%). *Anal.* Calcd. for $C_{14}H_{32}NiN_6O_4$: C, 41.29; H, 7.92; N, 20.64. Found: C, 41.23; H, 7.66; N, 21.24.

Preparation of Chelating Nitrite Complexes

1. [Co(2,6-dimethylpyridine *N*-oxide)₂(NO₂)₂]

First the chloride [Co(2,6-dimethylpyridine *N*-oxide)₂Cl₂] complex is prepared. 2,6-Dimethylpyridine *N*-oxide (2,6-lutidine *N*-oxide) (2.46 g., 0.02 mole) (Aldrich Chemical Co.) is

---

* To 0.04 mole of pyridine add 0.5 mole (11.5 g.) sodium in large pieces during 30 minutes. Add 50 ml. absolute ethanol and heat until the sodium disappears. Do not let cool! Begin to distill the alcohol and add 55–60 ml. water slowly at first and then as rapidly as possible to keep the solid portion to a minimum. After the alcohol is gone, there will be left a yellow oil layer and an aqueous layer. Extract the aqueous layer with ether. Dry the oil and extracts over potassium carbonate and evaporate the solvent.[9] 2-[(Methylamino)methyl]piperidine collected at 55° and 8 mm. is a colorless oil which turns yellow on contact with air.

added to a solution of 2.38 g. (0.01 mole) cobalt(II) chloride hexahydrate dissolved in a minimum amount of absolute methanol (8 ml.).   Ten milliliters of acetone is added, and the solution is kept in a refrigerator overnight.   The blue crystals are filtered by suction, washed several times with acetone, then dried *in vacuo*.   The compound decomposes between 264–265°. Yield is 2.89 g. (74%).

Very finely powdered silver nitrite is prepared by dissolution of silver nitrate (6 g.) and sodium nitrite (2.5 g.) in a minimum amount of water and mixing the resulting solutions.   Light yellow crystals of silver nitrite are removed by filtration.   Four and eight tenths grams (0.032 mole) of silver nitrite is added to the 2.89 g. (0.008 mole) of Co(2,6-dimethylpyridine *N*-oxide)$_2$Cl$_2$ in 800 ml. of dry acetone.   (Acetone is refluxed with anhydrous potassium carbonate and then distilled.)   The solution is stirred by a magnetic stirrer overnight (8 hours).   The blue solution of the cobalt(II) chloride complex turns mauve.   The solution is filtered by suction and then evaporated by vacuum until mauve crystals form.   The crystals are filtered by suction and dried *in vacuo*.   The compound decomposes between 215–219°.   Yield   is   2.4   g.   (83%).   *Anal.*   Calcd.   for C$_{14}$H$_{18}$CoN$_4$O$_6$: C, 42.32; H, 4.56; N, 14.10.   Found: C, 42.55; H, 4.92; N, 14.05.

2. [Co(2,4,6-trimethylpyridine *N*-oxide)$_2$(NO$_2$)$_2$]

Same procedure as in Procedure 1 on page 204.   Four and fourteen hundredths g. (0.03 mole) of 2,4,6-trimethylpyridine *N*-oxide [obtained by hydrogen peroxide oxidation* of 2,4,6-tri-methylpyridine (Eastman) by the method of Boekelheide and

---

* The pyridine was oxidized with hydrogen peroxide and acetic acid at 70°. The mixture was concentrated as far as possible at 80° and the residue made alkaline with anhydrous sodium carbonate.   The sodium acetate and sodium carbonate were filtered and the chloroform evaporated leaving the 2,4,6-trimethylpyridine *N*-oxide as a high boiling oil.

Linn[10]] is added to 3.57 g. (0.015 mole) cobalt(II) chloride hexahydrate in 15 ml. of methanol. Acetone (15 ml) is added, and the solution is allowed to stand overnight in a refrigerator. The crystals are filtered by suction, washed several times with acetone, and then dried *in vacuo*. The cobalt(II) chloride compound decomposes between 285 and 286°. The yield of blue crystals is 3.0 g. (47%).

Four and eight tenths grams (0.032 mole) of silver nitrite is added to the 3.0 g. (0.008 mole) of $Co(2,4,6$-trimethylpyridine N-oxide$)_2Cl_2$ in 800 ml. of dry acetone. The solution is then treated as in Procedure 3. The compound decomposes between 230 and 232°. The yield of the mauve crystals is 1.87 g. (61%). *Anal.* Calcd. for $C_{16}H_{22}CoN_4O_6$: C, 45.18; H, 5.21; N, 13.17. Found: C, 50.82; H, 3.94; N, 11.22.

3. [Co(4-methylquinoline N-oxide)₂(NO₂)₂]

The chloride complex is made by adding 4.89 g. (0.03 mole) of 4-methylquinoline N-oxide [obtained by hydrogen peroxide oxidation* of 4-methylquinoline (lepidine) (Pierce Chemical Company) by the method of Ochiai[11]] to a solution of 3.57 g. (0.015 mole) cobalt(II) chloride hexahydrate dissolved in 12 ml. of absolute ethanol. This solution is stirred and then evaporated at a mild temperature on a hot plate. The residue is recrystallized from absolute ethanol, filtered by suction, washed several times with cold ethanol, and then dried *in vacuo*. Dark blue crystals are obtained which melt between 220 and 222°. Yield is 5.54 g. (84%). Then add 4.8 g. (0.032 mole) of silver nitrite to 2.77 g. (0.008 mole) of $Co(4$-methylquinoline N-oxide$)_2$-Cl_2 in 800 ml. of dry acetone. The solution is treated as in Procedure 1 on page 204. Brown crystals are obtained from a reddish-brown solution. Yield is 1.15 g. (42%). *Anal.* Calcd. for $C_{20}H_{18}CoN_4O_6$: C, 51.18; H, 3.86; N, 11.94. Found: C, 50.82; H, 3.94; N, 11.22.

---

* Use the same procedure as the preceding footnote, except the product precipitates and should be recrystallized from ether (m.p. 118–119°).

Properties

The nitrite group has three fundamental vibrational modes which are all active in the infrared region, and upon coordination the band positions are shifted as compared to the free nitrite frequencies. In Table I the infrared frequencies of the NO_2^-

T A B L E I   **Infrared Frequencies of the NO_2^- Groups, cm.$^{-1}$**

Compound*	ν_{as}	ν_s	ν_b
NaNO₂†,‡	1328 ± 2	1261 ± 2	828
[Ni{2-[(methylamino)methyl]pyridine}₂(ONO)₂]	1375(s)	1170(s)	812, 828§
[Ni{2-[(methylamino)methyl]piperidine}₂(ONO)₂]	1372(b)	1200(s)	821, 838§
[Co(2,6-dimethylpyridine N-oxide)₂(NO₂)₂]	1278(ms)	1207(vs)	859(ms)
[Co(2,4,6-trimethylpyridine N-oxide)₂(NO₂)₂]	1292(ms)	1219(vs)	858(ms)
[Co(4-methylquinoline N-oxide)₂(NO₂)₂]	1281(sh)	1214(vs)	856(m)

* As Nujol and hexachlorobutadiene mulls.
† J. Chatt, L. A. Duncanson, B. M. Gatehouse, J. Lewis, R. S. Nyholm, M. L. Tobe, P. F. Todd, and L. M. Venanzi, *J. Chem. Soc.*, **1959**, 4073.
‡ R. E. Weston and T. T. Brodasky, *J. Chem. Phys.*, **27**, 683 (1957).
§ Assignment was uncertain owing to the presence of a ligand band in this region.

groups are listed for the nickel(II) and cobalt(II) complexes described here. Visible and near-infrared spectral data have also been reported for both sets of complexes.[6,8]

References

1. S. M. Jorgensen, *Z. Anorg. Chem.*, **5**, 147 (1893); **19**, 109 (1899).
2. E. Billmann, *Z. Anal. Chem.*, **39**, 284 (1900).
3. F. Basolo and G. S. Hammaker, *J. Am. Chem. Soc.*, **82**, 1001 (1960).
4. J. Chatt, L. A. Duncanson, B. M. Gatehouse, J. Lewis, R. S. Nyholm, M. L. Tobe, P. F. Todd, and L. M. Venanzi, *J. Chem. Soc.*, **1959**, 4073.
5. D. M. L. Goodgame and M. A. Hitchman, *Inorg. Chem.*, **4**, 721 (1965).
6. L. El-Sayed and R. O. Ragsdale, *ibid.*, **6**, 1644 (1967).
7. D. M. L. Goodgame and M. A. Hitchman, *ibid.*, **3**, 1389 (1964); **5**, 1303 (1966).
8. L. El-Sayed and R. O. Ragsdale, *ibid.*, **6**, 1640 (1967).
9. C. S. Marvel and W. A. Lazier, "Organic Syntheses," Collective Vol. 1, p. 93, John Wiley & Sons, Inc., New York, 1932.
10. V. Boekelheide and W. J. Linn, *J. Am. Chem. Soc.*, **76**, 1286 (1954).
11. E. Ochiai, *J. Org. Chem.*, **18**, 534 (1953).

42. RUTHENIUM AMMINES*

$$2RuCl_3 + Zn + 16NH_3 \rightarrow 2[Ru(NH_3)_6]Cl_2 + [Zn(NH_3)_4]Cl_2$$
$$[Ru(NH_3)_6]Cl_2 + [Zn(NH_3)_4]Cl_2 + 4HCl \rightarrow$$
$$[Ru(NH_3)_6][ZnCl_4] + 4NH_4Cl$$
$$2[Ru(NH_3)_6]Cl_2 + Br_2 + 4HCl \rightarrow$$
$$2[Ru(NH_3)_5Cl]Cl_2 + 2HBr + 2NH_4Cl$$
$$2[Ru(NH_3)_6]Cl_2 + Br_2 + 4NaBr \rightarrow 2[Ru(NH_3)_6]Br_3 + 4NaCl$$

Submitted by J. E. FERGUSSON† and J. L. LOVE†
Checked by J. N. ARMOR‡

The preparation of the hexaammine complexes of ruthenium(II) and ruthenium(III) salts are sketchily described in the literature. The preparation of hexaammineruthenium(II) by the reduction of ruthenium trichloride with zinc in ammonia is described briefly by Lever and Powell.[1]   Allen and Senoff[2] carry out the reduction using hydrazine hydrate.   The hexaammineruthenium(III) cation is obtained by oxidation of the ruthenium(II) complex,[1] and pentaamminechlororuthenium(III) dichloride is obtained by treating the former compound with hydrochloric acid.[1,3]   This compound may also be obtained by treating the pentaammine molecular nitrogen complex of ruthenium(II) with hydrochloric acid.[2,4]

These compounds are of interest as useful starting materials for further complexes of ruthenium(II) and (III)[5,7] and recently have become the center of more interest in their connection with the formation of ruthenium(II) molecular nitrogen complexes.

* Since this synthesis was submitted, a more detailed description of the preparation of hexaammineruthenium(II) dichloride has been reported.[7]

† University of Canterbury, Christchurch, New Zealand.   For the support of this work acknowledgment is made to the New Zealand Universities Grants Committee and Johnson Matthey Chemicals, Ltd., for the loan of ruthenium trichloride.   Appreciation is expressed to Mrs. C. Sligh for testing the experimental details.

‡ Stanford University, Stanford, Calif. 94305.

Procedure

A. HEXAAMMINERUTHENIUM(II) DICHLORIDE

$$2RuCl_3 + Zn + 16NH_3 \rightarrow 2[Ru(NH_3)_6]Cl_2 + [Zn(NH_3)_4]Cl_2$$

Commercial-grade ruthenium trichloride (5 g.) is "activated" by dissolving in concentrated hydrochloric acid (10–20 ml.) and then evaporating the solution to dryness at 100° on a steam bath. The chloride is dissolved in 0.880 ammonia (50 ml.), excess AR zinc dust (1 g.) added, and the solution boiled for 7 minutes (Note 1). Considerable effervescence occurs during the reaction; therefore a 250-ml. Erlenmeyer flask would be a satisfactory reaction vessel. The excess zinc is filtered off and any crystallized yellow product washed out from the zinc by dissolving in the minimum quantity of water. This solution is added to the filtrate. If any zinc hydroxide precipitates out in the filtrate during this process, a few drops of ammonia (0.880) are added to redissolve it. Small portions of solid ammonium chloride are added to the filtrate and dissolved until crystallization of the product $[Ru(NH_3)_6]Cl_2$ commences. The solution is cooled in ice until crystallization is complete, and the product is filtered and washed with a little cold 0.880 ammonia and then ethanol (Note 2). Yield is 3 g. (Note 4, below). *Anal.* Calcd. for $[Ru(NH_3)_6]Cl_2$: N, 30.7%. Found: N, 30.3%.

Notes

1. Ruthenium metal may be produced if the reduction is allowed to proceed for a longer time.

2. Solutions of $[Ru(NH_3)_6]^{2+}$ are subject to aerial oxidation, and isolation of the solid complex should be completed as rapidly as possible.

B. HEXAAMMINERUTHENIUM(II) TETRACHLOROZINCATE

$$[Ru(NH_3)_6]Cl_2 + [Zn(NH_3)_4]Cl_2 + 4HCl \rightarrow$$
$$[Ru(NH_3)_6][ZnCl_4] + 4NH_4Cl$$

The filtrate remaining after collecting $[Ru(NH_3)_6]Cl_2$ in the above preparation is just neutralized with concentrated hydrochloric acid (Note 3). Pale yellow leaflets of the product begin to separate. These are filtered off and washed with a little cold water and then ethanol. Yield is 2 g. (Note 4). *Anal.* Calcd. for $[Ru(NH_3)_6][ZnCl_4]$: N, 20.4%. Found: N, 19.8%.

Notes

3. If more acid is added than is required for neutralization, or if the solution becomes too hot, oxidation occurs to give a deep-blue solution containing $[Ru(NH_3)_3H_2OCl_2]Cl$.[1] The complex cation $[Ru(NH_3)_6]^{2+}$ can be reformed if this blue solution, after being made alkaline with ammonia solution, is again reduced with zinc dust.

4. All of the product of the reaction given in Sec. A above can be isolated as the tetrachlorozincate if zinc chloride (*ca.* 2 g.) is added to the cold solution after neutralization with acid. No ammonium chloride is added in this case. The ratio of yields of the two products $[Ru(NH_3)_6]Cl_2$ and $[Ru(NH_3)_6][ZnCl_4]$ may vary from preparation to preparation, but the total yield of $[Ru(NH_3)_6]^{2+}$ will remain constant. The variation appears related to the concentration of ammonia in Sec. A at the time of isolation of the product. The yield is increased by increasing the ammonia concentration.

C. PENTAAMMINECHLORORUTHENIUM(III) DICHLORIDE

$$2[Ru(NH_3)_6]Cl_2 + Br_2 + 4HCl \rightarrow$$
$$2[Ru(NH_3)_5Cl]Cl_2 + 2HBr + 2NH_4Cl$$

Hexaammineruthenium(II) dichloride (3 g.) (Note 5) is dissolved in water (20 ml.) and bromine water added carefully until there is a slight excess. This is observed by the yellow solution first becoming paler, almost colorless, and then *slightly orange* as excess bromine accumulates (Note 6). Concentrated hydrochloric acid is added to give a 1:1 acid solution, and the solution is heated under reflux for 2 hours; excess bromine is first allowed to escape. The product $[Ru(NH_3)_5Cl]Cl_2$ is formed as yellow-orange crystals during the heating, and these are collected and recrystallized from boiling 1 M hydrochloric acid. Yield is 3 g. *Anal.* Calcd. for $[Ru(NH_3)_5Cl]Cl_2$: N, 23.9%. Found: N, 24.0%.

Notes

5. $[Ru(NH_3)_6][ZnCl_4]$ may be used instead of $[Ru(NH_3)_6]Cl_2$ in this preparation (and for Sec. D below). In this case not all of the complex completely dissolves before adding the bromine water, but it dissolves during the reaction.

6. If bromine is not present in excess, then a deep-blue solution is obtained on addition of hydrochloric acid.[7] On heating, a brown-yellow precipitate forms which is mainly $[Ru(NH_3)_3Cl_3]$. If too large an excess of bromine is added, a trace of this compound is also produced.

D. HEXAAMMINERUTHENIUM(III) TRIBROMIDE

$$2[Ru(NH_3)_6]Cl_2 + Br_2 + 4NaBr \rightarrow 2[Ru(NH_3)_6]Br_3 + 4NaCl$$

Hexaammineruthenium(II) dichloride (1 g.) (see Note 5 above) is oxidized with bromine water, as described above in Sec. C. After oxidation solid sodium bromide is dissolved in the neutral solution until the product $[Ru(NH_3)_6]Br_3$ forms as a yellow powder. The product is collected and redissolved in water (5 ml.) and reprecipitated by the addition of sodium bromide. The product is washed finally with ethanol. Yield is 1 g. *Anal.* Calcd. for $[Ru(NH_3)_6]Br_3$: N, 19.0%. Found: N, 18.8%.

Analysis

Nitrogen is determined by the Kjeldahl method using Devarda's alloy. The complex and the alloy are placed in a standard Kjeldahl apparatus and the ammonia distilled off from a 7.5 M sodium hydroxide solution into 2% boric acid. The ammonia is titrated with standard hydrochloric acid using bromocresol green–methyl red as indicator.

Properties

The yellow hexaammineruthenium(II) cation is a moderately strong reducing agent and will, for example, reduce hydrogen chloride, mercuric chloride, gold(III) chloride,[1] and the hexaamminecobalt(III) cation.[6] The dry complexes are stable for a matter of weeks, especially when kept cold, but in aqueous solution decomposition is more rapid.

The yellow pentaammine complex of ruthenium(III) chloride crystallizes as octahedral crystals. It is a stable compound and exists in hydrochloric acid solution over a wide range of acid concentrations. The pale yellow hexaammine complex of ruthenium(III) is soluble in water and an excellent starting material for further ruthenium(III) compounds.[5]

The infrared spectra of the complexes have been recorded, and assignments have been made.[2] The Ru(II) and Ru(III) ammines are readily distinguished by their infrared absorptions around 1300 cm.$^{-1}$.

References

1. F. M. Lever and A. R. Powell, "International Conference on Coordination Chemistry," p. 135, special publication no. 13, The Chemical Society, London, 1959.
2. A. D. Allen and C. V. Senoff, *Can. J. Chem.*, **45**, 1337 (1967).
3. L. H. Voft, J. L. Katz, and S. E. Wiberley, *Inorg. Chem.*, **4**, 1158 (1965).
4. A. D. Allen, F. Bottomley, R. O. Harris, V. P. Reinsalu, and C. V. Senoff, *J. Am. Chem. Soc.*, **89**, 5595 (1967).

5. K. Gleu and K. Rehm, *Z. Anorg. Allgem. Chem.*, **227**, 237 (1936).
6. J. F. Endicott and H. Taube, *Inorg. Chem.*, **4**, 437 (1965).
7. F. M. Lever and A. R. Powell, *J. Chem. Soc.* (A), **1969**, 1477.

43. PENTAAMMINECHLORORHODIUM(III) DICHLORIDE AND PENTAAMMINEHYDRIDORHODIUM(III) SULFATE

$$\text{RhCl}_3 \cdot 3\text{H}_2\text{O} \xrightarrow{\text{NH}_3/\text{EtOH}} [\text{Rh}(\text{NH}_3)_5\text{Cl}]\text{Cl}_2 \xrightarrow[\text{SO}_4{}^{2-}]{\text{Zn}/\text{NH}_3} [\text{Rh}(\text{NH}_3)_5\text{H}]\text{SO}_4$$

Submitted by J. A. OSBORN,* K. THOMAS,* and G. WILKINSON*
Checked by H. M. NEUMANN†

The catalytic effect of several alcohols in the preparation of dichlorotetrakis(pyridine)rhodium(III) cation has long been known.[1] In recent years, a variety of reducing agents, present in catalytic quantities, have been used in the preparation of several rhodium(III) complexes.[2] In the absence of catalysts, these reactions are often laborious, and/or incomplete, by comparison with the catalyzed reaction, for example, the preparation of pentaamminechlororhodium(III) chloride (Claus' salt) by the method of Lebedinsky.[3] Conversion of $[\text{Rh}(\text{NH}_3)_5\text{Cl}]\text{Cl}_2$ to the pentaamminehydridorhodium(III) salt $[\text{Rh}(\text{NH}_3)_5\text{H}]\text{SO}_4$ by treatment with zinc and ammonia is rapid, and the reaction is relatively clean.[4] The formation of hydrido species by tetrahydroborate treatment[5] is not a satisfactory preparative procedure.

Procedure

A. PENTAAMMINECHLORORHODIUM(III) CHLORIDE

Commercial hydrated rhodium trichloride (5.2 g., 20 mmoles), water (100 ml.), and ethanol (20 ml.), all contained in a conical

* Inorganic Chemistry Laboratories, Imperial College, London, S.W. 7, England.
† Georgia Institute of Technology, Atlanta, Ga. 30332.

flask (250 ml.), are gently heated at 30° until complete solution is obtained. Ammonia solution (density 0.88, 20 ml.) is then added and the mixture gently swirled until the resulting fawn suspension is homogeneous. This is then heated rapidly to the boil, with constant agitation, whereupon the solution clears to a pale yellow color. On slow cooling to room temperature, large, bright yellow crystals are formed; for maximum yield the mixture is cooled further in ice. The crystals are collected, washed with acetone (20 ml.), and dried at 60°. Yield is 4.6 g. (80% based on $RhCl_3 \cdot 2H_2O$). Evaporation of the filtrate to 30–40 ml. gives a further crop of crystals. The overall yield is essentially quantitative. Recrystallization of either crop is unnecessary, but may be performed from the minimum of boiling water.

B. PENTAAMMINEHYDRIDORHODIUM(III) SULFATE

A suspension of the finely powdered complex $[Rh(NH_3)_5Cl]Cl_2$ (6 g.) and ammonium sulfate (9 g.) in a mixture of 50 ml. of water and 50 ml. of ammonia (density 0.88), is heated to 60°. Zinc dust (total 2 g.) is added in three equal portions, at ½-minute intervals and the resulting suspension maintained at 60° for a further 2 minutes. Particles of rhodium metal and undissolved zinc are screened off by passing through a fine, sintered-glass filter bed and the clear filtrate is saturated with ammonia gas, while being kept at <5° by cooling in ice. It is essential that the filtrate be agitated briskly during the ammonia saturation stage. The product, which separates as a chalk-white solid, is filtered off, washed with two aliquots of acetone (each 50 ml.), and dried in a vacuum.

The product reacts with oxygen; consequently the purification stage must be carried out under nitrogen. The crude product is dissolved in deoxygenated water (75 ml.) through which nitrogen is bubbled. The contaminant, which remains as a white flocculate, is removed by screening through filter paper. Ammonium sulfate (2 g.) is added to the screened solution and the product obtained as a pale creamy microcrystalline solid by saturating the ice-cooled solution with ammonia gas as before. The product is collected, washed with acetone, and dried in a vacuum. Yield is 3.2 g. {54% based on [Rh(NH$_3$)$_5$Cl]Cl}. For maximum yield, it is *essential* that the ice-cooled solution be saturated with ammonia; ammonia should be introduced at as fast a rate as is possible.

Properties

Crystalline pentaamminehydridorhodium(III) sulfate is quite stable in air and can be stored indefinitely. In aqueous solution there exists the equilibrium:

$$[Rh(NH_3)_5H]^{2+} + H_2O \rightleftarrows [Rh(NH_3)_4(H_2O)H]^{2+} + NH_3$$

The proton magnetic resonance spectrum has a doublet centered at τ 27.1 (J_{Rh-H}, 14.5 Hz.; the Rh—H stretching frequency, in the infrared spectrum (Nujol mull), is at 2079(s) cm.$^{-1}$. The interaction of the complex with alkenes produces[5] stable alkyl complexes of the type [Rh(NH$_3$)$_5$R]SO$_4$. In solution the complex reacts with molecular oxygen to give a blue peroxo complex; displacement of ammonia by ethylenediamine can also be achieved.[6]

References

1. M. Delépine, *Bull. Soc. Chim. France*, **45**, 235 (1929).
2. (a) R. D. Gillard, J. A. Osborn, and G. Wilkinson, *J. Chem. Soc.*, **1965**, 1951; (b) R. D. Gillard and G. Wilkinson, *Inorganic Syntheses*, **10**, 64 (1967).
3. W. W. Lebedinsky, *Izv. Inst. Izuceniju Platiny*, **13**, 9 (1936); see, S. A. Anderson and F. Basolo, *Inorganic Syntheses*, **7**, 214 (1963).
4. K. Thomas, T. S. Osborn, A. R. Powell, and G. Wilkinson, *J. Chem. Soc.* (A), **1968**, 1801.
5. J. A. Osborn, R. D. Gillard, and G. Wilkinson, *J. Chem. Soc.*, **1964**, 3168.
6. K. Thomas and G. Wilkinson, *J. Chem. Soc.* (A), **1970**(2), 356.

44. DICHLORO(ETHYLENEDIAMINE)PALLADIUM(II) AND (2,2'-BIPYRIDINE)DICHLOROPALLADIUM(II)

Submitted by B. JACK McCORMICK,* EDGAR N. JAYNES, JR.,* and ROY I. KAPLAN*
Checked by H. C. CLARK† and J. D. RUDDICK†

Dichloro(ethylenediamine)palladium(II) and (2,2'-bipyridine)-dichloropalladium(II) are useful intermediates for the preparation of mixed-ligand complexes. In compounds derived from these intermediates in which the amine ligands are preserved, effects of the different bonding properties of ethylenediamine and 2,2'-bipyridine may be observed. In contrast to the rather involved synthesis of [Pt(en)Cl$_2$],[1] the syntheses given here are straightforward and provide pure products in high yield. The procedures used are somewhat similar to those given elsewhere,[2-5] but are placed on a considerably more specific basis than those given hitherto.

Procedure

A. DICHLORO(ETHYLENEDIAMINE)PALLADIUM(II)

$$2K_2PdCl_4 + 2en \rightarrow [Pd(en)_2][PdCl_4] + 4KCl$$
$$[Pd(en)_2][PdCl_4] \xrightarrow[H^+]{\Delta} 2[Pd(en)Cl_2]$$

Five grams (0.0153 mole) potassium tetrachloropalladate(II) is dissolved in 40 ml. of warm water, and the resulting solution is divided into two equal portions. To one of the portions in a 125-ml. Erlenmeyer flask is added 10.5 ml. (0.0158 mole) of 1.50 M ethylenediamine. The ethylenediamine should be

* West Virginia University, Morgantown, W.Va. 26506.
† University of Western Ontario, London, Ontario.

added dropwise and with vigorous stirring. The resulting mixture, which contains [Pd(en)$_2$]Cl$_2$, [Pd(en)Cl$_2$], and [Pd-(en)$_2$][PdCl$_4$], is then heated to boiling for 1.5 hours. During the heating period the volume of the solution should be maintained at approximately 50 ml., and after the heating period any undissolved precipitate is removed by filtration.

The filtrate is cooled to 10°, and to the cool solution is added slowly and with stirring the remaining 20-ml. portion of K$_2$PdCl$_4$. The voluminous pink precipitate of [Pd(en)$_2$][PdCl$_4$] is separated by filtration and washed with several 10-ml. portions of cold water. The precipitate then is suspended in 60 ml. of water containing five drops of 6 *M* HCl; this mixture is evaporated on a steam bath to approximately 10 ml. During the evaporation the [Pd(en)$_2$][PdCl$_4$] is converted to yellow [Pd(en)Cl$_2$]. This conversion may be slow and the evaporation with dilute HCl should be repeated several times until the color change is complete. The mixture is then cooled in an ice bath and filtered. The yellow crystals are washed with several 10-ml. portions of cold water, air-dried, and then dried *in vacuo* over CaCl$_2$. Yield is 3.2–3.5 g. (89–97%, based on K$_2$PdCl$_4$). *Anal.* Calcd. for [Pd(en)Cl$_2$]: C, 10.1; H, 3.38; Pd, 44.8. Found: C, 10.3; H, 3.32; Pd, 44.7.

B. (2,2′-BIPYRIDINE)DICHLOROPALLADIUM(II)

$$2K_2PdCl_4 + 2bipy \rightarrow [Pd(bipy)_2][PdCl_4] + 4KCl$$
$$[Pd(bipy)_2][PdCl_4] \xrightarrow[H^+]{\Delta} 2[Pd(bipy)Cl_2]$$

To 1.19 g. (0.00763 mole) of 2,2′-bipyridine in 40 ml. of methanol is added slowly a solution of 2.50 g. (0.00767 mole) potassium tetrachloropalladate(II) in 40 ml. of water. The pale-pink to tan product of [Pd(bipy)$_2$][PdCl$_4$] is collected on a suction filter and washed thoroughly with 80 ml. of water to remove potassium chloride.

The moist product then is placed in a beaker and suspended in 60 ml. of water containing six drops of 6 M HCl. The mixture is heated on a steam bath to reduce the volume to 20 ml. After about one hour all of the [Pd(bipy)$_2$][PdCl$_4$] is converted to yellow [Pd(bipy)Cl$_2$]. The solution and solid are cooled in an ice bath and filtered. The light yellow complex thus isolated is washed with several 10-ml. portions of cold water, air-dried, and then dried *in vacuo* over calcium chloride. Yield is 2.2–2.5 g. (87–97%, based on K$_2$PdCl$_4$). *Anal.* Calcd. for [Pd(bipy)-Cl$_2$]: C, 36.0; H, 2.40; Pd, 31.9. Found: C, 35.7; H, 2.41; Pd, 32.1.

Properties

Dichloro(ethylenediamine)palladium(II) is formed as yellow needles. (2,2′-Bipyridine)dichloropalladium(II) is produced as a light yellow microcrystalline solid that can be obtained as larger crystals by careful recrystallization from hot, dilute HCl solutions. Both compounds are insoluble in cold water, but are moderately soluble in hot water. The compounds dissolve in 1 M sodium hydroxide with reaction. In general the compounds are not soluble in organic solvents; however, both complexes can be dissolved in dimethyl sulfoxide and N,N-dimethylformamide.

The NH$_2$-stretching frequencies are located at 3312 and 3220 cm.$^{-1}$, and the NH$_2$-deformation frequency is located at 1575 cm.$^{-1}$ in [Pd(en)Cl$_2$].

References

1. G. C. Johnson, *Inorganic Syntheses*, **8**, 242 (1966).
2. H. D. K. Drew, F. W. Pinkard, G. H. Preston, and W. Wardlaw, *J. Chem. Soc.*, **1932**, 1895.
3. D. E. Ryan, *Can. J. Res.*, **27B**, 938 (1949).
4. R. A. Walton, *Spectrochim. Acta*, **21**, 1795 (1965).
5. J. L. Burmeister and F. Basolo, *Inorg. Chem.*, **3**, 1587 (1964).

45. NONAHYDRIDORHENATE SALTS

Submitted by A. P. GINSBERG* and C. R. SPRINKLE*
Checked by J. F. RUSSELL,† F. N. TEBBE,† and E. L. MUETTERTIES†

Salts of $[ReH_9]^{2-}$ [1] and $[TcH_9]^{2-}$ [2] are so far the only examples of transition-metal hydride complexes in which there is no ligand other than hydrogen. Investigation of these compounds has been inhibited by the small yields and impure products obtained in the method of preparation originally described.[3] This difficulty has been overcome for $ReH_9{}^{2-}$ by a synthesis of the disodium salt[4] in which an ethanol solution of sodium perrhenate is reduced with sodium metal to give the hydride in *ca.* 35% yield. About 15% of the unconverted perrhenate is lost in side reactions, but the remainder may be recovered. Metathesis between $Na_2[ReH_9]$ and $(Et_4N)_2SO_4$ allows the preparation of $(Et_4N)_2[ReH_9]$, a salt which has proved very useful in studying the reactions of nonahydridorhenate with tertiary phosphines[5] and with carbon monoxide.[6]

A. DISODIUM NONAHYDRIDORHENATE

$$NaReO_4 \xrightarrow[\text{EtOH}]{xsNa} Na_2[ReH_9]\downarrow + NaOEt + NaOH‡$$

Procedure

A 500-ml., three-necked flask, equipped with a nitrogen inlet, a mechanical stirrer, and a West condenser, is charged with a filtered solution of sodium perrhenate§ (3.0 g., 11 mmoles)

* Bell Telephone Laboratories, Inc., Murray Hill, N.J. 07974.
† Central Research Department, Experimental Station, E. I. du Pont de Nemours & Company, Wilmington, Del. 19898.
‡ The stoichiometry of the reaction is uncertain.
§ It is less expensive to buy potassium perrhenate and convert it to the sodium salt by ion exchange. A solution of 10 g. of the potassium salt in 900 ml. water is passed through a 2.5 × 12-cm. column of 200–400 mesh AG50W-X12 cation exchange resin (Bio-Rad Laboratories) in the hydrogen form. The effluent is titrated with 1 N sodium hydroxide (carbonate-free) to a potentiometric end point; the neutral solution is evaporated to dryness. The product is crushed to a powder and oven-dried at 120°.

in absolute ethanol (300 ml.). As the solution is stirred rapidly and the flask flushed with a stream of nitrogen, sodium spheres* (2 g., 87 mmoles) are slowly added via the condenser. When the solution turns dark brown, it is heated to reflux with a heating mantle; the heat is turned off, and additional sodium spheres (10 g., 435 mmoles) are added at such a rate that the solution is kept refluxing but the capacity of the condenser is not exceeded. After the sodium has dissolved completely, the hot solution is centrifuged (2000 r.p.m., 5 minutes) and the supernatant is reserved for recovery of perrhenate. The precipitate is extracted with *ca.* 3% ethanolic sodium ethoxide† (two times with 25 ml., about one hour each time) and then washed successively with 2-propanol (three times with 25 ml.) and ether (three times with 25 ml.). The extraction and washing are most conveniently done by using the first portion of sodium ethoxide solution to transfer the precipitate from the large centrifuge bottle to a 40-ml., glass-capped, centrifuge tube and then shaking the suspension with a mechanical shaker. After preliminary drying in a nitrogen stream, the white product is pumped for one hour at $25°/10^{-3}$ mm. and then overnight at $82°/10^{-3}$ mm. Yields vary between 0.7 and 1.2 g. (26–45% based on the starting perrhenate, 47–91% based on the amount of unrecovered perrhenate); the average yield of a large number of runs was 0.92 g. (35% based on the starting perrhenate, 71% based on unrecovered perrhenate). The reaction may be run in about the same percent yield with all amounts doubled. The hydride should be stored in a dry, CO_2-free, inert atmosphere.

Infrared spectra of the dried product (4000–250 cm.$^{-1}$, KBr disk)‡ show weak impurity bands at *ca.* 3450 [ν(OH)], 2910, and

---

* Reagent grade, $\frac{1}{16}$ to $\frac{1}{4}$ in. diam., wash with pentane and clean by brief immersion in ethanol immediately before use.

† Prepared by dissolving sodium (15 g.) in absolute ethanol (250 ml.) under nitrogen and diluting to 500 g. in a polyethylene bottle.

‡ Samples for infrared spectroscopy should be prepared in a dry, inert-atmosphere box. It is best to protect the sample with layers of pressed KBr above and below the sample bearing layer.

2840 [ν(CH)], 2700 (?), *ca.* 1640 [δ(OH)]; 1450 [δ(CH)], and [ν_3($CO_3{}^{2-}$)], and 910–930 cm.$^{-1}$ [ν(ReO)]. Partial purification may be effected by the following procedure. All solvents and solutions are kept ice-cold and are deaerated with a nitrogen stream or by evacuating to boiling and back-filling with nitrogen. The procedure should be carried through rapidly and exposure of hydride solutions to air minimized by flushing open vessels with a nitrogen stream. The crude product (1.0 g.) is dissolved in carbonate-free 25% sodium hydroxide solution* (5 ml.) in a capped, 40-ml., polyethylene, centrifuge tube. Methanol (10 ml.) is added, and the mixture is centrifuged (10,000 r.p.m., 0–20°, 5 minutes).† The clear but colored supernatant is poured with stirring (magnetic stirrer) into 200 ml. of ethanol contained in a 250-ml. centrifuge bottle. The white precipitate which forms is centrifuged down (2000 r.p.m., 5 minutes) and transferred to a 40-ml. capped centrifuge tube with 25 ml. of fresh ethanol. After shaking for 15 minutes (prolonged contact with ethanol in the absence of base leads to decomposition), the ethanol is removed and the solid is extracted with 2-propanol (twice with 25 ml., 15 minutes each time) then washed with anhydrous ether, and dried as before to give 0.90 g. (90%) of a white powder. *Anal.* Calcd. for Na₂[ReH₉]: Na, 19.05; Re, 77.18; H, 3.76. Found: Na, 19.11; Re, 76.66; H, 3.76; C, 0.36. Weak impurity bands are still present in the infrared spectrum. (See Fig. 12.) Analysis of the crude product before reprecipitation gave the following results: Na, 18.22; Re, 75.82; H, 3.89; C, 0.44. The sample used for this analysis was from a mixture of the products of some 30 reactions; analysis of the crude product from individual reactions shows some variation.

---

* Prepared by diluting the clear supernatant from a centrifuged (10,000 r.p.m., 10 minutes) 50% solution. On mixing 5 ml. of the 25% solution with 10 ml. of methanol and 200 ml. of ethanol there should be no visible turbidity after one hour standing on ice.

† The 40-ml. tube may be kept sufficiently cold during centrifugation by fitting it into an ice-cold, massive, rubber adapter which fits the cavity of a 250-ml. angle centrifuge head.

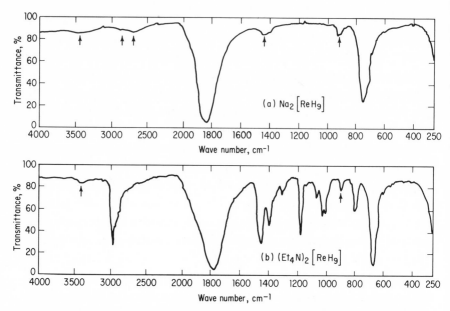

Fig. 12. *Infrared spectra (KBr sandwich) of* (a) Na₂[ReH₉] *and* (b) (Et₄N)₂-[ReH₉]. *The arrows point to impurity bands. These spectra may be used to judge the purity of the product obtained. If the relative intensities of the impurity bands do not exceed those shown, the material will have a satisfactory analysis.*

Unreacted perrhenate may be recovered from the reaction mixture by stirring in 5 ml. of 20% aqueous potassium hydroxide and allowing the solution to stand overnight. A precipitate of potassium perrhenate settles out.

Properties

The infrared spectrum (Nujol mull) of Na₂[ReH₉] has ν(ReH) at 1835(s), (br) and δ(ReH) at 745(s), *ca.* 720(sh), and 630(sh) cm.$^{-1}$. In aqueous alkali $\tau_{Re-H} = 19.1$. The compound is soluble in water and methanol, slightly soluble in ethanol, and insoluble in 2-propanol, acetonitrile, ether, and tetrahydrofuran. Alkali stabilizes the water and methanol solutions. With acids the hydride evolves hydrogen very vigorously. On heating in a vacuum (2–6°/minute) visible decomposition begins at *ca.* 245°. When a *ca.* 50-mg. sample of $Na_2[ReH_9]$ was exposed to the atmosphere at a relative humidity of 14% (25°), it turned light gray in about one hour, and an infrared spectrum showed an increase in the intensity of the $\nu_3(CO_3^{2-})$ and $\nu(ReO)$ impurity bands. After overnight exposure at 14% relative humidity, the sample turned black; an infrared spectrum indicated that rhenium hydride was still present, but $\nu_3(CO_3^{2-})$ and $\nu(ReO)$ were now strong bands and $\nu_2(CO_3^{2-})$ had appeared at 882 cm.$^{-1}$. In this experiment $\nu(OH)$ increased only moderately on exposure to the atmosphere, but under more humid conditions, decomposition is more rapid and $\nu(OH)$ becomes strong.

Analytical Methods

Nonahydridorhenate is conveniently oxidized to perrhenate by slow addition of the solid to 5% hydrogen peroxide. Sodium and rhenium may then be determined on the solution by standard methods. Hydrogen is determined by ordinary combustion techniques.

B. BIS(TETRAETHYLAMMONIUM) NONAHYDRIDORHENATE

$$Na_2ReH_9 + (Et_4N)_2SO_4 \xrightarrow[\text{H}_2\text{O-EtOH}]{\text{Et}_4\text{NOH}} Na_2SO_4\downarrow + (Et_4N)_2ReH_9$$

Procedure

All solvents and solutions are degassed; open flasks and funnels are flushed with a nitrogen stream, and evacuated flasks are back-filled with nitrogen after filtration or evaporation steps. Crude $Na_2[ReH_9]$ (1.10 g., 4.56 mmoles, *ca.* 10% more than the amount required for stoichiometric metathesis) is dissolved in an ice-cold bis(tetraethylammonium) sulfate solution of known concentration (0.332 N, 25.0 ml.) in 10% aqueous tetraethylammonium hydroxide,* contained in a 250-ml. Erlenmeyer flask. Ice-cold ethanol (150 ml.) is added and the mixture allowed to stand on ice for *ca.* 5 minutes, after which it is rapidly filtered through a medium-porosity fritted disk. The filtrate is evaporated to dryness on a rotary evaporator ($25°/10^{-2}$ mm., liquid-nitrogen trap), and the residue is redissolved in 2-propanol (50 ml.). After filtration through a fine-porosity fritted disk, the solution is evaporated to dryness as before. The process of dissolving the residue in 2-propanol and evaporating to dryness is repeated until the ν(OH) band (3400 cm.$^{-1}$) in the infrared spectrum of the solid (KBr sandwich, dry-box) is weak or negligible (see Fig. 12*b*); a total of two or three evaporations is usually sufficient.† The crude product (2.0 g., 100%) is white or slightly colored, and its infrared spectrum has a ν(ReO) band at 910 cm.$^{-1}$, as well as several other impurity bands; it is, however, suitable for most synthetic purposes and should be transferred and stored under dry nitrogen or argon.

Purification of the crude $(Et_4N)_2[ReH_9]$ may be effected by the following procedure, which is carried out in a dry-box using acetonitrile dried with calcium hydride and ether distilled from lithium tetrahydroaluminate. The product is dissolved in acetonitrile (40 ml.), and ether (50 ml.) is immediately added to precipitate a white or light tan solid‡ which is collected by

---

* To prepare a stock solution (500 ml.) mix appropriate stoichiometric amounts of standard 6 N sulfuric acid with standardized 10% aqueous Et₄NOH (Eastman), evaporate to near dryness (70°, 15 mm.), and dissolve the residue in 10% aqueous [Et₄N]OH to 500.0 ml.

† We have obtained the best results by carrying out the evaporations in rapid succession. A small amount of brown decomposition product usually forms during the final evaporation.

‡ Nonahydridorhenate reacts slowly with acetonitrile to give a brown product. If the solution becomes dark brown, the precipitate will be colored. The impurity is intensely colored, and the sample color is not a good criterion of analytical purity.

vacuum filtration on a coarse, fritted disk. After thoroughly washing with ether, the product is dried for 10 minutes in an argon stream and then for 8 hours at $56°/10^{-5}$ mm. (1.6 g., 80%). An infrared spectrum shows no significant impurity bands, although if the crude material contained an appreciable amount of tetraethylammonium hydroxide, the purification procedure does not remove it and a ν(OH) band will be present. *Anal.* Calcd. for $[(C_2H_5)_4N]_2[ReH_9]$: C, 42.15; H, 10.83; N, 6.14; Re, 40.87. Found: C, 41.82; H, 11.03; N, 5.86; Re, 40.73.

Properties

The infrared spectrum (Nujol mull) of $(Et_4N)_2ReH_9$ has ν(ReH) at 1780(s), (br) and δ(ReH) at *ca.* 720(sh), 675(s), and *ca.* 610(sh) cm.$^{-1}$. In acetonitrile solution, $\tau_{Re-H} = 18.5$. The compound is soluble in water, acetonitrile, ethanol, 2-propanol, and other alcohols; it is insoluble in ether, tetrahydrofuran, and 1,2-dimethoxyethane. The solutions are stabilized by alkali. On heating in a vacuum ($2-4°/$minute), decomposition occurs at 115–120° with the evolution of hydrogen and ethane. When a *ca.* 50-mg. sample of $(Et_4N)_2[ReH_9]$ was exposed to the atmosphere at a relative humidity of 19% (25°), it formed moist, brownish clumps within half an hour. An infrared spectrum taken after one hour exposure showed strong ν(ReO) and δ(OH) bands at 910 and 1630 cm.$^{-1}$, respectively; the ν(ReH) and δ(ReH) frequencies were present but diminished in intensity.

References

1. S. C. Abrahams, A. P. Ginsberg, and K. Knox, *Inorg. Chem.*, **3**, 558 (1964).
2. A. P. Ginsberg, *ibid.*, **3**, 567 (1964).
3. A. P. Ginsberg, J. M. Miller, and E. Koubek, *J. Am. Chem. Soc.*, **83**, 4909 (1961).
4. A. P. Ginsberg and C. R. Sprinkle, *Inorg. Chem.*, **8**, 2212 (1969).
5. A. P. Ginsberg, *Chem. Commun.*, **1968**, 857.
6. A. P. Ginsberg and M. J. Hawkes, *J. Am. Chem. Soc.*, **90**, 5930 (1968).

46. POTASSIUM HEXAKIS(ISOTHIOCYANATO)-NIOBATE(V)

$$NbCl_5 + CH_3CN \rightarrow NbCl_5 \cdot CH_3CN$$
$$NbCl_5 \cdot CH_3CN + 6KNCS \rightarrow K[Nb(NCS)_6] + 5KCl + CH_3CN$$

Submitted by G. F. KNOX* and T. M. BROWN*
Checked by W. A. G. GRAHAM† and G. O. EVANS†

A preponderance of the transition-metal pseudohalogen complexes reported in the literature are prepared in aqueous media. Several oxidation states of many transition metals are either unstable in the presence of water or form only oxygen-coordinated species. Thus, these metal ions will not form pseudohalogen complexes in the normal manner. The following method, using polar, nonaqueous solvents is suitable for the preparation of isothiocyanate complexes of several of these ions. As an example of the preparation of such complexes, the synthesis of potassium hexakis(isothiocyanato)niobate(V) is described.

Niobium pentachloride decomposes readily in the presence of moisture; thus it is very important in this synthesis that the acetonitrile be rigorously dried. The method described by Coetzee *et al.*[1] is suitable for the purification of this solvent. The niobium pentachloride starting material should also be freed of any oxy species. This can be effected at about 100° by sublimation of the pentachloride away from the less volatile oxychlorides using a standard vacuum sublimer with a water-cooled probe. Appearance of commercial samples is a poor guide to their purity, and this purification step should not be omitted. The potassium thiocyanate should be purified by recrystallization and then thoroughly dried.

* Arizona State University, Department of Chemistry, Tempe, Ariz. 85281.
† The University of Alberta, Edmonton 7, Canada.

Procedure

The apparatus is shown in Fig. 13. The air and moisture sensitivity of Nb(V), in both the starting material and final product, dictates the use of inert atmosphere and vacuum-line techniques for this preparation. The Teflon needle-valve stopcocks employed in the apparatus allow a wide variety of solvents to be used, as no stopcock grease is present in the system. The vessel is degassed thoroughly before the synthesis is initiated. Note that the stirring bars must be inserted into flasks A and C prior to making the constrictions in side arms B and D.

In an inert atmosphere, 2.70 g. (0.01 mole) of niobium pentachloride is placed in reaction bulb C through side arm D, and

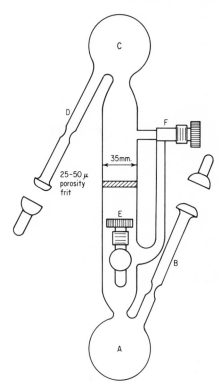

Fig. 13. Apparatus for the preparation of potassium hexakis(isothiocyanato)niobate(V).

5.83 g. (0.06 mole) of potassium thiocyanate is placed in reaction bulb A through side arm B along with a magnetic stirring bar. The two side arms are capped, and the reaction vessel is removed and placed on a vacuum manifold. After evacuation to approximately 10^{-5} torr, side arms B and D are sealed off with a torch. Fifty to sixty milliliters of acetonitrile is then vacuum-distilled into the vessel by immersing bulb A in a cold bath. When sufficient solvent has been introduced into the flask, Teflon stopcock E is closed, the vessel removed from the manifold, and the solvent allowed to reach ambient temperature. Reaction bulb A is then immersed in a warm water bath (60–70°) and stirring initiated. Teflon stopcock F is opened to allow passage of solvent vapors into the upper chamber through the side arm. The vapors condense in the upper chamber and react with the pentachloride to form the highly soluble monoacetonitrile adduct. This solution then passes through the glass frit and flows into the potassium thiocyanate solution in chamber A. An immediate reaction occurs, as evidenced by the dark red color formed in the lower solution. To ensure completeness of reaction, this stirring and heating of the lower chamber is continued for about 2 hours after all of the niobium pentachloride has been extracted from the upper chamber.

Teflon stopcock F is now closed and the solvent vapor forced to pass through the frit into the upper chamber where it condenses. After about one-half of the solvent has thus been distilled into the upper chamber, the vessel is inverted, and the highly soluble K[Nb(NCS)$_6$] passes into chamber C, while the relatively insoluble potassium chloride is retained by the frit. Even though potassium chloride is quite insoluble in acetonitrile, it shows appreciable solubility in the solution of K[Nb(NCS)$_6$] due to the increased ionic strength of the solvent. Thus, the distillation of some of the solvent into the upper chamber before filtration reduces the amount of potassium chloride contamination in the final product.

The vessel is now returned to the vacuum manifold and the

acetonitrile solvent removed by vacuum distillation. Care should be taken during the distillation, for the product tends to form a solid layer over the liquid and then "spatter" over the inside of the vessel as the vapor pressure of the solvent breaks this layer. When all of the liquid is removed, the acetonitrile of solvation is removed by placing the chamber containing the product in a water bath maintained at approximately 80° and keeping the product under a dynamic vacuum of 10^{-5} torr for 48–72 hours. Teflon stopcock E is then closed, and the reaction vessel removed to the dry-box.

In the dry-box, Teflon stopcock E is opened and bulb C (containing the crude product) is removed from the vessel by scoring and breaking the neck. The deep blue crystalline product (slightly contaminated with 1–2% KCl) is removed and stored. (Yield is *ca.* 90–95%.)

Purification of the crude product is effected in a vessel similar to that in Fig. 13, but side arm B may be deleted. The product is placed in chamber C, the side arm capped, and the vessel evacuated. After side arm D is sealed off, about 60–70 ml. of dry, degassed 1,2-dichloroethane is distilled into chamber A and Teflon stopcock E closed. The slightly soluble $K[Nb(NCS)_6]$ is then extracted from the insoluble potassium chloride by the 1,2-dichloroethane using the same techniques discussed earlier for extraction of the niobium pentachloride by acetonitrile. Due to the low solubility of the complex in the solvent, the extraction requires about 5–6 days to extract 2.5–3.0 g. of product. During the dichloroethane extraction, there is a tendency for the frit to plug, and care must be taken that the solution above the frit does not overflow, carrying potassium chloride down through the side arm into the extracted product. The rate of filtration can be increased by occasionally shutting the stopcock on the side arm, so that dichloroethane vapor is forced up through the frit, agitating the residue. When purification is complete, the solvent is removed, and the product vacuum-dried in the manner previously described for the crude product in acetonitrile solvent. The K[Nb(NCS)$_6$] must be handled in a dry-box. Final yield is 3.5–4.0 g. (74–83%). *Anal.* Calcd. for K[Nb(NCS)$_6$]: K, 8.14; Nb, 19.34; N, 17.49; C, 15.00; S, 40.04. Found: K, 8.44; Nb, 19.06; N, 17.25; C, 15.10; S, 39.79.

Properties

Potassium hexakis(isothiocyanato)niobate(V) is a lustrous blue, crystalline solid. It is very soluble in acetonitrile, giving a red-to-maroon solution. It is slightly soluble in 1,2-dichloro-ethane, dichloromethane, and chloroform, whereas it decomposes in acetone and other oxygen-containing solvents. It decomposes in the presence of moisture or in water solution, with the evolution of hydrogen sulfide. In acetonitrile, it reacts with quaternary amine and tetraphenylarsonium chloride salts to yield potassium chloride and the hexakis(isothiocyanate) complex of the respective cation. It does not melt or show signs of decomposition below 250°. X-ray powder photographs indicate it is isostructural with the potassium hexakis(isothio-cyanato)tantalate(V). The infrared mull spectrum has four bands in the 1900–2100-cm.$^{-1}$ range, but in acetonitrile solvent, the complex has only two bands, a strong one at 1981 cm.$^{-1}$ and a weak one at 2027 cm.$^{-1}$, in this region.

Discussion

The synthetic method for the preparation of thiocyanate complexes of air- and water-unstable transition-metal oxidation states has been found suitable for the synthesis of K[Ta(NCS)$_6$] and K$_2$[M(NCS)$_6$] (where M = Ti^{4+},[2] (Zr^{4+}, Nb^{4+}, Mo^{4+}, or W^{4+}), as well as the compound described above. By use of different solvents (e.g., diethyl ether or nitromethane), it can be expanded to include metal halides which undergo reduction in the presence of acetonitrile (e.g., $MoCl_5$ and WCl_5). Thus, by use of the proper solvents, many thiocyanate complexes, and possibly other pseudohalogen complexes, can be prepared in this manner.

References

1. J. F. Coetzee, G. P. Cunningham, D. K. McGuire, and G. R. Padmanabhan, *Anal. Chem.*, **34**, 1139 (1962).
2. O. Schmitz-DuMont and B. Ross, *Z. Anorg. Allem. Chem.*, **342**, 82 (1966).

Corrections

DIAQUAHYDROGEN *trans*-DICHLOROBIS-(ETHYLENEDIAMINE)COBALT(III) DICHLORIDE*

Submitted by JACK M. WILLIAMS†

The compound formulated as *trans*-dichlorobis(ethylenediamine) cobalt(III) chloride in Volume 2^1 has been found to contain the diaquohydrogen ion^2 $(H_5O_2)^+$ rather than "adduct" hydrogen chloride alone as implied from the first equation on page 223. The $(H_5O_2)^+$ ion is present as $(H_2O \cdot H \cdot OH_2)^+$ and not as $(H_3O^+ \cdot H_2O)^{2-4}$. The first equation on page 223 of Volume **2** should more appropriately read:

$$4CoCl_2 + 8C_2H_4(NH_2)_2 + 8HCl + 6H_2O + O_2 \rightarrow$$
$$4[H(H_2O)_2][trans\text{-}Co(en)_2Cl_2]Cl_2$$

Both water and hydrogen chloride are released simultaneously when the bright green crystalline plates are dried at 110°, in static air, leaving dull green polycrystalline $[Co(en)_2Cl_2]Cl$.

References

1. J. C. Bailar, *Inorganic Syntheses*, **2**, 222 (1946).
2. J. M. Williams, *Inorg. Nucl. Chem. Letters*, **3**, 297 (1967). This name chosen in accordance with the 1960 I.U.P.A.C. definitive rules for inorganic chemistry.

* Research supported by the U.S. Atomic Energy Commission.
† Argonne National Laboratory, Argonne, Ill. 60439.

3. R. D. Gillard and G. Wilkinson, *J. Chem. Soc.*, **1964**, 1640.
4. J. M. Williams in "Molecular Dynamics and Structure of Solids," R. S. Carter and J. J. Rush (eds.), special publication no. 301, National Bureau of Standards, Washington, D.C., **1969**, p. 237.

TRIS(ETHYLENEDIAMINE)CHROMIUM(III) SULFATE: A MODIFIED PROCEDURE

$$Cr_2(SO_4)_3 + 6C_2H_4(NH_2)_2 \rightarrow 2[Cr(en)_3]_2(SO_4)_3$$

Submitted by W. N. SHEPARD*
Checked by F. BASOLO† and M. NICOLINI†

The following modification of the procedure, previously published [*Inorganic Syntheses*, **2**, 198 (1946)], is recommended.

A 500-ml. round-bottomed flask, to which a water-cooled condenser is attached by a ground joint, is used to reflux 49 g. of $Cr_2(SO_4)_3 \cdot 18H_2O$, which has been ground to a powder, and 50 ml. of anhydrous ethylenediamine in a heating mantle. After about one hour, the product, orange-yellow, is usually formed. A spatula is used to scrape the compound off the inner wall of the flask and to mix it thoroughly. The product is then placed in a hot-water bath for 3–4 hours to remove unreacted ethylenediamine. The solid is broken up with a spatula, ground, washed with alcohol, and air-dried. Yield is 89 g. [95%, based on $Cr_2(SO_4)_3 \cdot 18H_2O$].

* University of Arizona, Tucson, Ariz. 85721.
† Northwestern University, Evanston, Ill. 60201.

INDEX OF CONTRIBUTORS

SUBJECT INDEX

Names used in this cumulative Subject Index for Volumes XI, XII and XIII, as well as in the text, are based for the most part upon the "Definitive Rules for Nomenclature of Inorganic Chemistry," 1957 Report of the Commission on the Nomenclature of Inorganic Chemistry of the International Union of Pure and Applied Chemistry, Butterworths Scientific Publications, London, 1959; American version, *J. Am. Chem. Soc.*, **82**, 5523–5544 (1960); and the latest revisions (in press as a Second Edition (1970) of the Definitive Rules for Nomenclature of Inorganic Chemistry); also on the Tentative Rules of Organic Chemistry—Section D; and "The Nomenclature of Boron Compounds" [Committee on Inorganic Nomenclature, Division of Inorganic Chemistry, American Chemical Society, published in *Inorganic Chemistry*, **7**, 1945 (1968) as tentative rules following approval by the Council of the ACS]. All of these rules have been approved by the ACS Committee on Nomenclature. Conformity with approved organic usage is also one of the aims of the nomenclature used here.

In line, to some extent, with *Chemical Abstracts* practice, more or less inverted forms are used for many entries, with the substituents or ligands given in alphabetical order (even though they may not be in the text); for example, derivatives of arsine, phosphine, silane, germane, and the like; organic compounds; metal alkyls, aryls, 1,3-diketone and other derivatives and relatively simple specific coordination complexes: *Iron, cyclopentadienyl-* (also at *Ferrocene); Cobalt(II), bis (2,4-pentanedionato)-* [instead of *Cobalt(II) acetylacetonate*]. In this way, or by the use of formulas, many entries beginning with numerical prefixes are avoided; thus, *Vanadate(III), tetrachloro-*. Numerical and some other prefixes are also avoided by restricting entries to group headings where possible: *Sulfur imides*, with the formulas; *Molybdenum carbonyl*, $Mo(CO)_6$; both *Perxenate*, $HXeO_6^{3-}$, and *Xenate-(VIII)*, $HXeO_6^{3-}$. In cases where the cation (or anion) is of little or no significance in comparison with the emphasis given to the anion (or cation), one ion has been omitted; e.g., also with less well-known complex anions (or cations): $CsB_{10}H_{12}CH$ is entered only as *Carbaundecaborate(1−), tridecahydro-* (and as $B_{10}CH_{13}^-$ in the Formula Index).

Under general headings such as *Cobalt(III) complexes* and *Ammines*, used for grouping coordination complexes of similar types having names considered unsuitable for individual headings, formulas or names of specific compounds are not usually given. Hence it is imperative to consult the Formula Index for entries for specific complexes.

Two entries are made for compounds having two cations and for addition compounds of two components, with extra entries or cross references for synonyms. Unsatisfactory or special trivial names that have been retained for want of better ones or as synonyms are placed in quotation marks.

Boldface type is used to indicate individual preparations described in detail, whether for numbered syntheses or for intermediate products (in the latter case, usually without stating the purpose of the preparation). Group headings, as *Xenon fluorides*, are in lightface type unless all the formulas under them are boldfaced.

As in *Chemical Abstracts* indexes, headings that are phrases are alphabetized straight through, letter by letter, not word by word, whereas inverted headings are alphabetized first as far as the comma and then by the inverted part of the name. Stock Roman numerals and Ewens-Bassett Arabic numbers with charges are ignored in alphabetizing unless two or more names are otherwise the same. Footnotes are indicated by *n.* following the page number.

FORMULA INDEX

The Formula Index, as well as the Subject Index, is a cumulative index for Volumes XI, XII and XIII. The chief aim of this index, like that of other formula indexes, is to help in locating specific compounds or ions, or even groups of compounds, that might not be easily found in the Subject Index, or in the case of many coordination complexes are to be found only as general entries in the Subject Index. *All* specific compounds, or in some cases ions, with definite formulas (or even a few less definite) are entered in this index or noted under a related compound, whether entered specifically in the Subject Index or not. As in the latter index, **boldface type** is used for formulas of compounds or ions whose preparations are described in detail, in at least one of the references cited for a given formula.

Wherever it seemed best, formulas have been entered in their usual form (*i.e.*, as used in the text) for easy recognition: Si_2H_6, XeO_3, NOBr. However, for the less simple compounds, including coordination complexes, the significant or central atom has been placed first in the formula in order to throw together as many related compounds as possible. This procedure often involves placing the cation last as being of relatively minor interest (*e.g.*, alkali and alkaline earth metals), or dropping it altogether: MnO_4Ba; $Mo(CN)_8K_4 \cdot 2H_2O$; $Co(C_5H_7O_2)_3Na$; $B_{12}H_{12}{}^{2-}$. Where there may be almost equal interest in two or more parts of a formula, two or more entries have been made: Fe_2O_4Ni and $NiFe_2O_4$; $NH(SO_2F)_2$, $(SO_2F)_2NH$, and $(FSO_2)_2NH$ (halogens other than fluorine are entered only under the other elements or groups in most cases); $(B_{10}CH_{11})_2Ni^{2-}$ and $Ni(B_{10}CH_{11})_2{}^{2-}$.

Formulas for organic compounds are structural or semistructural so far as feasible: $CH_3COCH(NHCH_3)CH_3$. Consideration has been given to probable interest for inorganic chemists, *i.e.*, any element other than carbon, hydrogen, or oxygen in an organic molecule is given priority in the formula if only one entry is made, or equal rating if more than one entry: only $Co(C_5H_7O_2)_2$, but $AsO(+)$-$C_4H_4O_6Na$ and $(+)$-$C_4H_4O_6AsONa$. Names are given only where the formula for an organic compound, ligand, or radical may not be self-evident, but not for frequently occurring relatively simple ones like C_5H_5 (cyclopentadienyl), $C_5H_7O_2$ (2,4-pentanedionato), C_6H_{11}(cyclohexyl), C_5H_5N(pyridine). A few abbreviations for ligands used in the text are retained here for simplicity and are alphabetized as such: "en" (under "e") stands for ethylenediamine, "py" for pyridine, "bipy" for bipyridine, "pn" for 1,2-propanediamine (propylenediamine), "fod" for 1,1,1,2,-2,3,3-heptafluoro-7,7-dimethyl-4,6-octanedionato, "thd" for 2,2,6,6-tetramethylheptane-3,5-dionato, "DH" for dimethylglyoximato and "D" for the dianion, $(CH_3)_2C_2N_2O_2{}^{2-}$.

The formulas are listed alphabetically by atoms or by groups (considered as units) and then according to the number of each in turn in the formula rather than by total number of atoms of each element. This system results in arrangements such as the following:

<table>
<tr><td>NHS₇</td><td>(FSO₂)₂NH (instead of F₂S₂O₄NH)</td></tr>
</table>

NHS_7 $(FSO_2)_2NH$ (instead of $F_2S_2O_4NH$)
$(NH)_2S_6$ (instead of $N_2H_2S_6$) FSO_3H
$NH_3B_{10}CH_{12}$ F_2SO_3

$[Mo(CO)_3C_5H_5]K$
$[Mo(CO)_3C_5H_4CH_3]$
                                          $[Cr(en)_3][Ni(CN)_5]$ ["en" instead of
                                          $(NH_2)_2C_2H_4$ or $N_2H_4C_2H_4]$
FNO $[Cr(NH_3)_6][Ni(CN)_5]$

Footnotes are indicated by n. following the page number.